ECONOMIC AND ENVIRONMENTAL RISK AND UNCERTAINTY

THEORY AND DECISION LIBRARY

General Editors: W. Leinfellner (*Vienna*) and G. Eberlein (*Munich*)

Series A: Philosophy and Methodology of the Social Sciences

Series B: Mathematical and Statistical Methods

Series C: Game Theory, Mathematical Programming and Operations Research

Series D: System Theory, Knowledge Engineering and Problem Solving

SERIES B: MATHEMATICAL AND STATISTICAL METHODS

VOLUME 35

Scope: The series focuses on the application of methods and ideas of logic, mathematics and statistics to the social sciences. In particular, formal treatment of social phenomena, the analysis of decision making, information theory and problems of inference will be central themes of this part of the library. Besides theoretical results, empirical investigations and the testing of theoretical models of real world problems will be subjects of interest. In addition to emphasizing interdisciplinary communication, the series will seek to support the rapid dissemination of recent results.

The titles published in this series are listed at the end of this volume.

ECONOMIC AND ENVIRONMENTAL RISK AND UNCERTAINTY

New Models and Methods

edited by

ROBERT NAU

Duke University, Fuqua School of Business,
Durham, North Carolina, U.S.A.

ERIK GRØNN

Norwegian School of Management, Department of Business Economics,
Sandvika, Norway

MARK MACHINA

University of California at San Diego, Department of Economics,
La Jolla, California, U.S.A.

and

OLVAR BERGLAND

Agricultural University of Norway, Department of Economics and Social Sciences,
Aas, Norway

KLUWER ACADEMIC PUBLISHERS

DORDRECHT / BOSTON / LONDON

A C.I.P. Catalogue record for this book is available from the Library of Congress.

ISBN 0-7923-4556-8

Published by Kluwer Academic Publishers,
P.O. Box 17, 3300 AA Dordrecht, The Netherlands.

Sold and distributed in the U.S.A. and Canada
by Kluwer Academic Publishers,
101 Philip Drive, Norwell, MA 02061, U.S.A.

In all other countries, sold and distributed
by Kluwer Academic Publishers Group,
P.O. Box 322, 3300 AH Dordrecht, The Netherlands.

Printed on acid-free paper

Printed in the Netherlands

Dedicated to the memory of OLE HAGEN
1921-1994

TABLE OF CONTENTS

PREFACE

Since the early 1980's, the biannual international conferences on Foundations and Applications of Utility, Risk, and Decision Theory (the "FUR" conferences) have provided a forum in which much of the most innovative contemporary research on risk and uncertainty has been presented. This volume contains a selection of the best papers presented at the FUR VII conference, which was organized by the late Ole Hagen and held in Oslo from June 30 to July 3, 1994. The theme of the FUR VII conference was "Environmental Risk," which is reflected in the title and organization of the volume: Part I contains papers dealing with health and environmental applications of risk theory, including the plenary address given at the conference by Graham Loomes. Part II contains papers dealing with more general issues of risk and uncertainty in economic models of auctions, markets, and games. Part III contains papers which focus on the modeling of preferences and risk attitudes within individuals. As Professor Wenstøp observes in his memorial to Hagen on the following pages, the FUR conferences have chronicled an exciting period in the history of utility and risk theory, during which many revolutionary ideas have been proposed, debated, and applied to the study of economic institutions and individual decision-making. The papers in this volume continue that distinguished tradition.

The editing of this volume has been a truly collaborative effort between Erik Grønn (Norwegian School of Management), Mark Machina (University of California at San Diego), Olvar Bergland (Agricultural University of Norway), and myself. Special thanks are due to Erik Grønn, who chaired the National Organizing Committee for the conference and who, at the last minute, also inherited the Program Chair's duties from Professor Hagen. He handled the difficult early work of collecting and organizing the many manscripts that were submitted for consideration in the volume, from which the papers presented here were later selected through a reviewing process that spanned four universities. The FUR VII conference was made possible through the generous financial support of the Norwegian School of Management and contributions from the Central Bank of Norway and the Norwegian Trade Council.

Robert Nau
Durham, North Carolina
August 14, 1996

OLE HAGEN AND THE FUR CONFERENCES

Volume 21 of the Theory and Decision Library, entitled *Expected Utility and the Allais Paradox,* was edited by Maurice Allais and Ole Hagen in 1979. Their intent was to initiate a debate concerning the foundations of risk and utility theory, and they were indeed successful. Volume 21 contained contributions by many distinguished authors, including Oskar Morgenstern, and addressed the conflict between what was then called the American School (Expected Utility Theory) and the French school (Allais' Positive Theory of Choice). That volume provided the inspiration for the FUR conferences. The idea was due to Werner Leinfellner, who was co-editor of the Theory and Decision Library. As a follow-up to the volume, he proposed that Ole Hagen arrange an international conference on the foundations of utility and risk theory.

Hagen had performed several empirical studies demonstrating that people do not in fact behave according to the von Neumann-Morgenstern theory. To account for the observed behavior, he had in the years 1969-1972 developed an axiomatic theory of choice which generalizes the standard theory in that the third moment of the probability distribution over utilities is taken into account as well. The theory was certainly not welcomed in all quarters, but as it turned out to be similar to Allais', Hagen and Allais became allies in an endeavor to instigate an open debate. Leinfellner's encouragement was therefore a welcome opportunity for Hagen to create a forum where both proponents of new ideas and defenders of the established paradigm could participate.

As a consequence, the first FUR conference took place at the Norwegian School of Management in Oslo in 1982. Hagen received enthusiastic local support during the preparations, notably at the University of Oslo where Bernt Stigum was instrumental. The conference attracted about 100 participants from around the world, and turned out to be a success by all standards. The debate turned decisively from whether new theories are at all required, to which ones are the most suitable under different conditions. By that time, there was mounting evidence from behavioral experiments about violations of the von Neumann-Morgenstern axioms--particularly the independence axiom. Several alternative axiomatic utility theories had recently been proposed to account for such behavior, and their authors were also present at the meeting. The contrasts among the new theories, the earlier theories of Allais and Hagen, and the standard theory of expected utility generated many exciting discussions. A revolution in "non-expected utility theory" had begun, and subsequent conferences in the FUR series would track the course of this revolution through the 1980's and into the 1990's.

It is no secret that the early debate appeared more like a combat than an exchange of ideas between scientists. Hagen was no exception: he felt he had to pound very hard on the door of established science to make people listen and consequently developed a very personal, almost polemic writing style to achieve it. For Hagen, the first FUR conference turned out to be a personal triumph when his many years of opposition to the established theory were finally vindicated. He did not rest on his laurels, however, but participated with keen interest in the ensuing series of conferences in Venice (1994),

Aix-en-Provence (1986), Budapest (1988), Durham (1990), and Cachan (1992). The conference series returned to Oslo as FUR VII in 1994, with Hagen as the chief organizer. We therefore received with sorrow the message of his sudden and untimely death on Friday, May 13, 1994, only weeks before the conference was to commence. Hagen was 73 years old on January 22 of that year.

Hagen came to the Norwegian School of Management in 1973 after having vainly advocated his unorthodox ideas in utility and risk theory at the Norwegian School of Economics and Business Administration in Bergen. His paper "Separation of Cardinal Utility and Specific Utility of Risk in the Theory of Choices under Uncertainty" in *Statsøkonomisk Tidsskrift* (Oslo 1969) gives an early presentation of his ideas. Later contributions are to be found in edited volumes from the series of FUR conferences.

To those who knew him, Hagen's mind seemed to work a bit differently from the ordinary. While scientists commonly tend to look for rules and regularities in their research, Hagen was always looking for exceptions. He was in fact bored by the regular, and missed no opportunity to pursue the irregular. We all know how frustrating exceptions can be, and Hagen's rather personal writing style probably contributed to the unusual intensity of the expected utility theory debate in the early years. His colleagues remember Hagen as an unusually friendly and soft-spoken person, however. He was knowledgeable in a broad sense, very perceptive, and an inexhaustible source of histories and anecdotes which could throw new light on any subject and enliven any discussion.

We miss Hagen as a friend and colleague and inspiring fellow researcher.

Fred Wenstøp
Oslo, June 12, 1996

I. Models of Environmental Risk

VALUING HEALTH AND SAFETY: SOME ECONOMIC AND PSYCHOLOGICAL ISSUES

MICHAEL JONES-LEE AND GRAHAM LOOMES
Department of Economics
University of Newcastle
Newcastle NE1 7RU, United Kingdom

Abstract

When making decisions about the allocation of scarce social resources, how should we weigh the benefits (or disbenefits) of changes in the health or safety of members of the population? The conventional (economic) wisdom is that the preferences of the affected population should be elicited - in the form of their willingness to trade off wealth for health/safety - and should then be incorporated into some overall cost-benefit analysis.

However, attempts to undertake such elicitations have run into a number of theoretical and practical problems. The purpose of this paper is to consider some of these issues and stimulate further thought and discussion about what they may signify and how they might be further explored.

Introduction

Let us start with a brief portrait of what standard economic theory tends to regard as a typical member of the population. Such an individual is assumed to have reasonably stable, well-specified preferences over more or less anything and everything s/he might have occasion to choose between or experience, including a very large array of private and public goods and services of all kinds, many of which will have some potential health and/or safety implications.

Given such comprehensive and well-behaved preferences, it should be possible to obtain estimates of the values that such individuals place on marginal changes in health or safety, and combine these values with those of other representatives of the relevant population to arrive at aggregate measures which can be incorporated into public policy and expenditure plans. Various ways of trying to obtain such values have been explored (for a succinct review, see Marin (1993)), but direct elicitation via some form of interview or questionnaire has been (and continues to be) widely advocated[1].

R. Nau et al. (eds.), Economic and Environmental Risk and Uncertainty, 3–32.

The challenge facing the designer of such a questionnaire is to formulate questions worded sufficiently clearly and unambiguously, and targeted sufficiently carefully, that they elicit precise, reliable and unbiased estimates of exactly the data that are of interest. Thus, for example, a representative sample of individuals might be asked to consider the possibility of purchasing a safety feature for their car which would reduce their risk of being killed in a road accident during the next twelve months from a current baseline of 8 in 100,000 down to 4 in 100,000. The data from such a sample can then be combined to give a 'value of a statistical life'[2] which may be used by the relevant public body when appraising any projects with possible implications for the number of deaths in road accidents.

This is a special case of an approach that has many other applications. Besides being used to value changes in the risks of premature death, it can in principle also be used to value changes in the risks of *non*-fatal road injuries. And if injuries from road accidents are amenable to such treatment, why not injuries from other causes such as work, defective products, sport and leisure activities, crime, etc.? And if we can value the (prevention of) damage to health and quality of life through injury, why not do likewise in the context of illness/disease? And why not also value the (prevention of) damage to quality of life from other sources such as congestion and environmental pollution? And so on.

Unfortunately, experience has shown us that there are a number of very substantial practical difficulties with this kind of approach. To begin with, there are a good many 'nuts and bolts' problems to do with finding forms of question that adequately convey the necessary information while at the same time remaining sufficiently clear and simple to be accessible to as broad a cross-section of the population as possible. Anyone who has been involved at first hand in the often lengthy process of designing, piloting, redesigning, repiloting, ..., will fully appreciate the extent and degree of these difficulties. However, such issues - important though they are - will not be our concern in this paper. In what follows we shall assume that the nitty-gritty problems of wording and presentation of information have been resolved, and concentrate instead on a series of issues that seem to be of an even more fundamental and intractable kind.

Given the constraints of time and space, the issues we shall focus on will certainly not constitute an exhaustive list. Moreover, our selection is strongly influenced by recent projects in which we have been involved and tends to reflect the principal concerns that arose in the course of those projects. Even so, we hope that the examples we shall discuss will serve to raise a number of more general and fundamental questions which have implications across a very broad field of human activity and social welfare. But first, in order to put what follows in context, some brief background information may be helpful.

Background

In 1989, having recently adopted a willingness-to-pay based value for the prevention of fatal road injuries[3], the U.K. Department of Transport began a process of commissioning research to determine corresponding values for the prevention of *non*-fatal road accident injuries. However, whereas the notion of a fatality is relatively well-defined, the official definition of a 'serious' non-fatal road injury is very broad, ranging from temporary conditions which may pass very quickly through to permanent damage involving severe disability and/or distress, including some outcomes which many people may regard as being as bad or worse than death.

It was therefore necessary to devise a set of 'typical' injury descriptions which between them would span the spectrum of serious injuries. In order to keep things manageable, the questions asking for direct monetary valuations focused on a subset of these injuries - labelled R, S, X and W - which together could be regarded as accounting for a large proportion of the total number of serious injuries. The prognoses relating to these four injuries are reproduced in Appendix A: each gives information about the period of time normally spent in hospital, and the level and duration of pain and disability likely to be experienced initially and subsequently. To provide a link with earlier research concerning fatalities, death (labelled K) was also included in the valuation exercise.

In all versions of all questionnaires, the first substantive task facing respondents was to read through the full set of prognoses and rank them from best (or least bad) to worst. In some phases of piloting, in the Main Study, and in follow-up work afterwards, respondents were also asked to locate the injuries on a single hundred-point visual analogue scale (VAS), with instructions encouraging them to use the distances between injuries on the scale to indicate the extent that they felt one prognosis was worse than another. Although a number of health state measurement exercises have used VAS scores to compute values[4], that was not our intention: for us, the primary purpose of the initial ranking and scoring exercise was to familiarise respondents with the prognoses and encourage them to begin to think about their relative severity.

Having completed that initial stage, respondents were provided with some additional information about the risks relating to each of the various injuries. Thus for each injury, they were presented with a **risk reduction scenario**, which gave them the current baseline annual risk of sustaining that injury, and then asked them to think about a safety feature which would reduce their own risk of incurring that injury by some specified amount.

We anticipated that when asked to state the value they placed on small changes in already-small risks of injury or death, many respondents might find it difficult to provide an instant point estimate which would accurately and reliably express their preferences. For this reason, we first asked them to place upper and lower bounds on the 'personal confidence interval' within which they felt sure their value lay. That is to say, we asked them about sums of money they were *certain they would pay*, and then noted the point at which they started to become uncertain as the amount increased: this point constituted the lower bound value, henceforth denoted by **min**. They were also asked to identify larger amounts they were certain they *would not* be prepared to pay, and then we noted the point at which they became less sure as the sums of money decreased: this constituted the upper bound, denoted by **max**. It was only after these personal upper and lower bounds had been elicited that each respondent was asked to try to pinpoint the particular value at which they would find the decision about whether or not to pay for the risk reduction most finely balanced[5], which was subsequently taken to be the **best** estimate of the individual's marginal rate of substitution of wealth for safety.

Piloting began in 1990. Of course, by that date there already existed a great deal of evidence from other sources - in particular, numerous surveys to elicit the value of (preventing) environmental damage - suggesting that people were liable to give responses which did not conform well with standard preference theory. We therefore decided to try out various forms of questions, and in each case to build in a number of internal consistency checks.

We were open to the possibility that people might have really quite wide personal confidence intervals; that they might find it difficult to identify a single **best** value within each interval; and that even when they did provide a single value, they might not have very great confidence in that value being a better indicator of their preferences than a number of adjacent values. Nevertheless, we still had some hopes that, on average, individuals' 'true' values would lie *somewhere* inside the stated intervals.

In the event, such hopes proved to be seriously overoptimistic. It became apparent from an early stage in the piloting of the questionnaire that there was more than just a little local imprecision in people's responses. And as piloting progressed and evolved into the Main Study, the problems manifested themselves in many diverse forms. Let us now consider some of those problems in more detail.

Same Scenarios, Same Response Mode, Different Cues

In the first phase of piloting, two versions of contingent valuation (CV) questionnaire were used[6]. These differed only in respect of the specified life-span of the vehicle safety feature: in one case, the safety feature lasted for the entire life of the respondent's vehicle, so that the value elicited was the lump sum the respondent would be willing to pay for such a feature; whereas in the other case, the safety feature had a one year life-span, so that the value elicited was the sum the respondent would be willing to pay to achieve a reduction in risk for the next 12 months[7].

In both questionnaires, the procedure for asking the willingness-to-pay questions was as follows. For each of the five prognoses (K, R, S, X and W) in turn, respondents were asked to look at a showcard which depicted the relevant risk reduction scenario. Alongside the showcard, the interviewer placed a disk mounted on a stand. The face of the disk was plain white, except for a window cut at the "twelve o'clock" position. In that window, a sum of money was displayed. Half of each version of the questionnaire specified that all valuation questions should begin with £25 displayed in the window of the disk, while the other half began each question with £75 displayed. For each scenario, the interviewer began by asking whether the respondent would or would not be willing to pay the sum displayed in the window; and then, depending on the response, rotated the disk to increase or decrease the sum, until the **min**, **max** and **best** estimates had been identified.

Thus within each of the two versions of CV questionnaire we can make five comparisons between pairs of subsample means, giving a total of ten comparisons overall. The null hypothesis is that the amount initially displayed makes no significant difference to the values subsequently elicited; while the alternative hypothesis is that values are significantly higher for the subsamples presented with the higher initial amount.

The data reveal a strong starting point effect: the null hypothesis of no significant effect was rejected at the 1% level in six cases, at the 5% level in three more cases, and at the 10% level in the tenth case - in all cases, in favour of the alternative hypothesis that the £75 starting point resulted in higher responses than the £25 starting point. Overall, the ratios of mean responses ranged from 1.89:1 to 2.87:1.

Although not unprecedented nor entirely unexpected[8], the fact that the amount initially displayed on the disk can have such a marked effect on the level of responses clearly raises doubts about the assumption that respondents have reasonably robust preferences which can be readily accessed by carefully constructed questionnaires.

One possible fall-back position is to concede that people's preferences *are* somewhat imprecise, and that they may therefore respond to these kinds of questions by simply drawing a value from their personal confidence interval. Thus even though there may be no overall difference between respondents' intervals, it might be that the initial stimulus - in this case, the first value displayed on the disk - could have an influence upon *whereabouts in the interval* they draw their response from. Given sufficiently wide intervals, the result may be to generate subsample mean values that appear to be significantly affected by the initial stimulus.

However, when we examine the various intervals, we find no support for such an explanation. On the contrary, in *every one* of the ten comparisons, the mean **min** for those initially shown £75 is substantially higher than the mean **max** for those initially shown £25, with the former being between 35% and 114% higher than the latter. Such disparities between *whole intervals*, induced by such an apparently irrelevant consideration as the 'asking price' initially presented by the interviewer, suggest that, rather than simply providing an initial stimulus for respondents to reveal their existing true values, the opening bid might actually represent a significant - perhaps even the principal - influence upon their responses[9].

In subsequent piloting, and in the Main Study that followed, the disks were replaced by payment cards - that is, each risk reduction scenario was printed on a separate page alongside a range of values. Respondents were given pencils and erasers, and for each risk reduction in turn were asked to put a tick (✓) next to each amount they were sure they *would* pay for a safety feature providing that risk reduction for one year, a cross (X) next to each amount they were sure they would *not* pay, and an asterisk (*) next to the price at which they would find it most difficult to decide whether or not to buy the safety feature. The greatest ticked amount and the smallest crossed amount were taken to be, respectively, the **min** and **max** of the individual's personal confidence interval, and the asterisk was taken to signify the **best** estimate of the value of the risk reduction to that individual.

In this second phase of piloting we tested for 'range effects': for half of the sample, the values printed on each page ranged from £0 to £500, plus a "more than £500" option; whereas for the other half, the range was £0 to £1500, plus "more than £1500", with most of the items in this latter range being approximately three times bigger than the amounts occupying corresponding positions in the former range.

The different ranges did not have *quite* such dramatic effects on stated values as the different initial amounts displayed on the disks. That is to say, in ten comparisons between subsample means, the £0-1500 range produced higher mean **best** estimates in only nine cases, and mean **min** values from the higher range subsamples exceeded mean **max** values from the lower range subsamples in only five of the ten comparisons. Of course, although this may seem rather tame by comparison with the disk effects, it is still a very striking departure from what we should expect from respondents with robust preferences of the kind conventionally assumed.

Same Scenarios, Same Response Mode, Different Angles

During the past fifteen years there have been numerous papers reporting what Knetsch and Sinden (1984) referred to as an "unexpected disparity" between the amount many individuals are willing to pay (WTP) for some marginal benefit and the amount they would be willing to accept (WTA) as compensation for a corresponding marginal disbenefit. In the context of health and safety, substantial disparities between the sums people state that they are willing to pay for marginal reductions in risks of death or injury and the amounts they say they would be willing to accept as compensation for tolerating marginally higher risks have been reported by Thaler (1980), Jones-Lee et al (1985), Viscusi et al (1987) and McDaniels (1992), among others.

One interpretation of such disparities is that they are due to what has variously been called the **endowment effect**, or **loss aversion**, or **status quo bias** (see Kahneman, Knetsch and Thaler (1991)). The basic idea is that, rather than evaluating outcomes in absolute terms as if from some neutral standpoint, people tend to code them as gains or losses relative to some reference point, and then treat them asymmetrically, with perceived losses looming larger than corresponding gains.

However, others have proposed alternative explanations for the data. For example, Brookshire and Coursey (1987) and Coursey et al (1987) have suggested that the disparities are principally due to inexperience and lack of incentives to reveal true considered preferences, arguing that if elicitation procedures were designed so as to provide adequate experience, feedback and incentives, the differences between WTP and WTA responses would reduce to levels consistent with standard theory. More recently, Hanemann (1991) has suggested that even quite substantial disparities may be entirely consistent with standard preference theory in cases where there are only very imperfect substitutes for the good being valued - and he cites health as one possible example of such a case.

The DoT pilot studies generated some data relevant to this debate. In addition to the five questions asking for respondents' WTP, the first two phases of piloting both included a further two WTA questions involving increases in the risks of injuries S and X equal in magnitude to the risk reductions respondents had considered just a minute or two earlier. As in the WTP exercise, the WTA procedure elicited both upper and lower bounds of personal confidence intervals as well as best estimates.

One possible explanation for any disparity between WTP and WTA might simply be that people's valuations are somewhat imprecise, and that when asked what they are willing to *pay*, they tend to pick a value from the lower end of their personal confidence interval; whereas in the face of a willingness to *accept* question, they opt for a value from the upper end of that same (possibly quite wide) interval.

And indeed, there was some evidence that this was happening. But much more striking was the result that for a substantial proportion of respondents there was *no overlap whatsoever* between the intervals produced by the WTA questions and those generated by the WTP questions: that is, out of the 152 individuals who gave usable responses in the first two phases of piloting, **min** WTA strictly exceeded **max** WTP for 52% in the case of injury S and for 56% in the case of injury X[10].

Such disparities - not just between point estimates but between whole intervals - again suggest that something more than simple imprecision is at work. Informal debriefing of some of the participants gave some insights into what may have been happening. When asked how much they would be willing to pay for the risk reduction, a number of respondents appeared to focus on the fact that the baseline risk was already pretty small, so that even if they took *no* extra precautions they felt that they were very unlikely to have an accident anyway; and thinking in this way, they were often reluctant to contemplate paying very large sums to increase survival probability by what seemed to be such a relatively small amount. By contrast, the willingness to accept question tended to shift their focus of attention: now they were more likely to think in terms of how they would feel if they *did* sustain an injury, knowing that there was a real possibility that the injury had been sustained as a result of accepting extra risk in return for money. From this perspective, the risk change seemed relatively large and had stronger overtones of personal culpability, and it therefore appeared quite natural to demand much greater compensation for the extra risk - or even to say that they would refuse to accept the extra risk at any price.

This kind of account of WTP-WTA disparities may seem to many to be rather more persuasive than attempts such as Hanemann's to reconcile WTP-WTA disparities within standard preference theory[11]. But this raises important questions both for analysis and for policy. Should the kind of thinking outlined above be regarded as reflecting non-standard preferences which are nevertheless in some sense stable and 'true' - and which should therefore be incorporated into formal models (e.g. Tversky and Kahneman's (1991) Reference Dependent Utility model)? Or should we in some way feed back disparate WTP and WTA responses to the individuals who generated them and ask them to consider further the extent to which they want to maintain some distinction between the two (encouraging them to make explicit their grounds for any such distinction)?

Clearly there are a number of difficulties and dangers here. On the one hand, there is the danger of accepting too easily the lightly-considered instant reactions of people who themselves may attach no very great significance to their (hypothetical) responses to somewhat abstract questions, but whose responses are then used to inform policy with very real life and safety implications. On the other hand, there is the danger of leading/forcing people's responses to conform with some particular preconceptions about what "rational" preferences ought to entail.

However, these difficulties extend far beyond starting point/range effects and WTP-WTA disparities (important though they are), and we shall postpone further discussion until we have added some further items to the catalogue of concerns.

Same Health States, Same Response Mode, Different Risk Reductions

In the first phase of piloting, the same five risk reduction scenarios were used for both versions of CV questionnaire. However, the results aroused a suspicion that responses were rather insensitive to some of the key elements in those scenarios, particularly the *magnitudes* of risk reduction. We tested for this in the second phase: besides using two different payment card ranges, as described earlier, we also used two somewhat different sets of risk reductions. Thus respondents were randomised between four Groups. Table 1 below summarises the key features of this 2 x 2 design.

Table 1: Phase 2 Pilot Study Design

	Risk Reductions (out of 100,000)	Payment Range
	K, R, S, X, W	£
Group A	4, 6, 12, 18, 10	0 - 1500
Group B	4, 6, 12, 18, 10	0 - 500
Group C	4, 4, 4, 12, 12	0 - 500
Group D	4, 4, 4, 12, 12	0 - 1500

As we saw earlier, both the starting point effect in the first phase, and the range effect in the second phase, were powerful and pervasive. Yet this very pervasiveness offers a possible way of adjusting for it: if the effect of a cue is to induce a systematic increase or decrease in the valuations of all of the risk reductions being considered, one means of ameliorating the impact of these cues is to adjust the individual's responses for each of the non-fatal injuries by dividing through by that individual's valuation of the reduction in risk of a fatal injury.

As a first step in this process, let us consider all 92 respondents who provided a complete set of **best** estimates of WTP for all five risk reduction scenarios. Dividing through by the valuations for the reduction of risk of death does indeed appear to remove significant range effects[12], and Table 2 below shows the results of pooling Group A with Group B, and also Group C with Group D, and calculating mean ratios: that is, each individual's WTP for each non-fatal risk reduction (of whatever magnitude) is divided by his/her WTP for the 4 in 100,000 reduction in the risk of death, and these ratios are then averaged across the subsample.

Table 2: Means and Standard Deviations of Relative WTP

	R/K	S/K	X/K	W/K
Groups A & B	1.14 (0.74)	0.98 (0.86)	0.63 (0.40)	0.52 (0.39)
Groups C & D	1.16 (0.73)	0.87 (0.52)	0.65 (0.33)	0.58 (0.39)

Table 2 shows that once responses have been adjusted for range effects in this way, there are no significant differences between the two sets of relative valuations, *despite the fact that the magnitudes of risk reductions varied quite substantially between the two pairs of Groups.* Thus even though Groups A & B were valuing a reduction in the risk of injury S that was *three* times the size of the reduction offered to Groups C & D, their average adjusted response was only one eighth bigger. And in two other cases - R and X - where Groups A & B were presented with a risk reduction that was 50% bigger, their mean adjusted responses were actually slightly smaller than those given by Groups C & D. In other words, respondents exhibited considerable insensitivity to the various differences in the magnitudes of risk reduction.

These results pose a potentially very serious problem. Recall that the objective of this study was to find some way of bringing the value of preventing a non-fatal injury into line with the value already adopted by the DoT for the prevention of a fatality. Thus the aim was to estimate the marginal rate of substitution of wealth for risk of a 'typical' serious injury I - denoted by m_I - *relative to* the marginal rate of substitution of wealth for risk of death - denoted by m_K.

A basic assumption underpinning the CV method of estimating the value of preventing a statistical death/injury is that when the changes in risk are sufficiently small, the value of m_I/m_K can be adequately approximated by:

$$\frac{WTP(\Delta p_I) \cdot \Delta p_K}{WTP(\Delta p_K) \cdot \Delta p_I}$$

where Δp_I and Δp_K are, respectively, the magnitudes of risk reduction for injuries I and K, and $WTP(\Delta p_I)$ and $WTP(\Delta p_K)$ are respondents' stated willingness to pay for those risk reductions. Clearly, if WTP responses are relatively insensitive to the sizes of Δp_I and Δp_K, the resulting estimates may be varied considerably simply by arbitrarily altering those magnitudes.

For example, since the risk reductions presented to Groups C & D for both K and S were the same (4 in 100,000), the estimate of m_S/m_K based on their responses would simply be 0.87, as shown in Table 2; however, since Groups A & B were presented with a risk reduction of 12 in 100,000 for S, the corresponding estimate for m_S/m_K based on *their* responses would be one-third of the raw value shown in Table 2, i.e. 0.33. Thus the identical prognosis S is imputed a value 2.66 times higher in one case than in the other, largely as a result of the size of risk reduction that happened to be selected by the researchers.

The previous paragraph illustrates the problem with an example of between-sample disparities. However, there is also worrying evidence of *within-subject* anomalies. Consider the responses of members of Groups A & B to the risk reductions for injuries X and W. Since the members of these Groups were presented with risk reductions of 18 in 100,000 and 10 in 100,000 for those two injuries, their respective sample mean values of m_X/m_K and m_W/m_K would be computed as 0.14 and 0.21. Now since the great majority of respondents ranked X as worse than W, it is clearly anomalous for them to generate a value for preventing W which is *half as high again* as their value for preventing X. Moreover, this result cannot be ascribed to a few outliers distorting the sample means: computations at the individual level give $m_W/m_K > m_X/m_K$ for no fewer than 37 of the 46 respondents in Groups A & B[13].

In short, given the apparent insensitivity of many respondents to the risk information presented to them, striking anomalies such as the one involving W and X will be easy to reproduce, and this casts serious doubt on the extent to which we can regard any particular values of m_I/m_K derived from CV surveys as being indicative of the 'true' preferences that public policy makers might wish to incorporate into their decision processes.

Same Risk Reductions, Same Response Mode, Different Health States

Besides the magnitude of the risk reduction itself, the other key component of each scenario was the health state prognosis conditional upon sustaining a particular type of injury. As mentioned earlier, the first substantive task in every questionnaire required each respondent to rank the various prognoses from best to worst. Ties were allowed, but occurred very rarely - in the vast majority of cases people gave strict orderings, with a high degree of concordance between different individuals' rankings; and in cases where they were asked to locate the prognoses on a visual analogue scale, it was very common for people to put at least five percentage points between prognoses that were adjacent in their rankings (and for the subset W, X, S, R and K, the distances were generally a good deal larger than that).

This evidence shows unambiguously that people *could and did* discriminate clearly between different prognoses when asked to make direct comparisons. However, their CV responses frequently suggested considerable insensitivity to the severity and/or duration of the different prognoses.

To begin with, consider Groups C & D in the second phase of piloting. As reported in the previous section, the study design involved the same risk reductions for R and S, and also the same reductions for X and W. In the latter case, 7 of the 46 individuals actually said that they would pay more to reduce the risk of W by 12 in 100,000 than to reduce the risk of X by the same amount; and a further 17 gave exactly the *same* response to both questions[14]. What is more, there was a similar frequency of identical responses for the questions concerning R and S - 15 out of 46. Some of these observations involved the same individuals in both cases, but others did not, so that altogether more than half of the respondents in Groups C and D (24 out of 46) either treated R and S the same, or else treated X and W the same, or both. This additional source of insensitivity to the CV questions, combined with the apparent insensitivity to the sizes of the risk reductions, heightened concern about the reliance that could be placed on the results produced by this method of eliciting people's preferences.

A third phase of piloting experimented with additional ways of encouraging people to discriminate more clearly between the various scenarios. A sixth scenario - offering a 12 in 100,000 reduction in the risk of S - was added to the five that had been presented to Groups C and D, and after initially ranking the various prognoses, respondents were asked to prioritise[15] the six scenarios before being invited to state their WTP for each one in turn. By asking the WTP questions in the order which corresponded to the modal pattern of prioritisation, by using only two magnitudes of risk reduction (4 and 12 in 100,000), and by applying both levels of risk reduction to injury S and asking those two WTP questions consecutively, we intended to give the CV method its best chance of producing responses that were sensitive to the two key elements in each scenario.

However, even under these apparently favourable conditions, a considerable degree of insensitivity was observed with respect both to the size of risk reduction and to the severity of the prognosis. In the light of this and the evidence from the earlier phases of piloting, it was decided not to base the Main Study exclusively on CV, as had been done in 1982 in the case of fatalities. Instead, the national sample was to be randomised between a CV questionnaire of the kind described above and an alternative Standard Gamble (SG) procedure which had also been tried out in the first phase of piloting. More will be said about the contrast between these two procedures in the next section. But first, let us consider some of the data from the CV half of the Main Study which shows that, despite trying to offer respondents every opportunity (short of leading them) to give discriminating responses, the patterns established during piloting were confirmed on a within-subject basis.

The insensitivity to different magnitudes of risk reduction was demonstrated by the fact that out of 395 individuals providing usable responses to both questions, fewer than half (180, or 45.5%) stated a *strictly higher* WTP for the 12 in 100,000 reduction in the risk of S than for the 4 in 100,000 reduction offered in the next question. Almost as many - 178 - gave the *same* response to both questions[16] (and another 37 stated a smaller WTP for the larger reduction). Thus when the sample mean m_S/m_K was calculated on the basis of the responses to the first question, it was estimated to be 0.26; whereas on the basis of the second question, that same ratio was estimated to be 0.64 - a figure nearly two and a half times bigger. This within-subject disparity is only fractionally smaller than the between-sample disparity reported in the previous section.

The insensitivity of the CV responses to differences in the prognoses can be seen by contrasting the direct rankings/scalings with the relativities implied by the WTP amounts. On a hundred point scale, the mean score for W was 71.4 compared with 57.4 for X (the medians were, respectively, 75 and 60) with nearly 89% of respondents (342 out of 385) ranking W as strictly better than X; yet the WTP responses show 136 of those 342 giving the *same* figures in both cases (with a further 64 actually saying that they would pay more to reduce the risk of W). Thus in contrast to a fourteen percentage point difference in mean VAS scores for the two injuries, the CV-based estimates for m_W/m_K and m_X/m_K differ by only two percentage points. Moreover, although the difference between mean and median VAS scores for S and R was even bigger, with 360 out of 387 ranking S as strictly better than R, it was still the case that 131 of those 360 stated the same WTP for both risk reduction scenarios (with a further 38 saying that they were willing to pay more to reduce the risk of S). It was in the light of these figures, and a number of others like them, that we recommended that the Department of Transport should effectively ignore the CV-based estimates and instead base its value for preventing a non-fatal road injury entirely on the alternative SG procedure used with the other half of the sample[17]. This was a recommendation that the DoT subsequently accepted.

But before discussing some of the possible explanations/implications of such response mode insensitivities, let us first consider the contrast with the results produced by the SG procedure - a contrast which may contribute to such a discussion.

Same Prognoses, Different Response Modes

In the Main Study, both the CV and the SG questionnaires began by asking respondents first to rank the various prognoses and then to locate them on the hundred-point VAS. But then, instead of asking them to prioritise and place a WTP value on the six risk reduction scenarios, the SG sample were presented with a series of questions in the following format (where Q denotes the particular injury description being evaluated relative to normal health and death):

> *Suppose you were in a road accident and you were taken to hospital. The doctors tell you that if you are treated in the usual way, you will certainly experience the consequences shown below on card Q. However, they also tell you that there is a different treatment available, but its outcome is not certain. If it succeeds you will be restored to your normal state of health. But if it fails you will die.*[18]

On the facing page were two columns, one showing the chances of success, running from "99 in 100" at the top to "10 in 100" at the bottom, with the other column showing the corresponding chances of failure. In addition, there were options allowing respondents to say that they would accept the risky treatment only if the chances of failure were less than 1 in 100; that they would not accept the treatment at any risk of failure; or, at the other end of the columns, that they would prefer death to injury Q.

Analogous to the CV procedure of identifying upper and lower bounds on their personal confidence intervals, respondents were first asked to tick all cases where they were confident they *would choose to accept* the risky treatment and cross all those cases where they were confident they *would not accept* the risky treatment Finally, an asterisk was to be placed against the case where they felt that the accept/reject decision was *most finely balanced.* Taking the asterisk to signify the respondent's point of indifference between the certainty and the gamble, and denoting the asterisked probability of failure by p, it can be shown that, to a good approximation, $p = m_Q/m_K$.

On the same basis, the largest ticked probability of failure can be regarded as an estimate of the respondent's **min**, while the smallest crossed probability of failure is taken as an estimate of **max**.

Table 3 contrasts the mean **best** estimates for the various m_Q/m_K ratios obtained from the CV and SG samples in the Main Study.

Table 3: Main Study Results

	CV	SG
m_R/m_K	0.875	0.233
$m_{S(4)}/m_K$	0.640	0.151
$m_{S(12)}/m_K$	0.262	
m_X/m_K	0.232	0.055
m_W/m_K	0.210	0.020
$m_{I(12)}/m_K$	0.352	0.117

Notes: $m_{S(4)}/m_K$ and $m_{S(12)}/m_K$ are estimated, respectively, on the basis of risk reductions of 4 in 100,000 and 12 in 100,000. The overall ratio between a serious non-fatal injury and death, $m_{I(12)}/m_K$, is a weighted average of the various m_Q/m_K, with the weights given by the relative frequencies of the different injuries, and using $m_{S(12)}/m_K$ rather than $m_{S(4)}/m_K$ in the calculation.

Table 3 shows that the CV questionnaire generated estimates of m_R/m_K and $m_{S(4)}/m_K$ which are *about four times higher* than the comparable SG-based estimates. What is more, as the severity of the injury decreases, the disparity appears to widen, so that even though the CV estimate of m_W/m_K is based on a risk reduction of 12 in 100,000 (which would thus tend to deflate it), it is still *more than ten times bigger* than the corresponding figure estimated from SG responses. The overall effect is that even when the lower CV-based value of m_I/m_K is taken, it is more than three times bigger than the SG-based figure.

Moreover, just as we saw in previous sections, the disparities are not simply a matter of differences between means. Examination of the personal confidence intervals shows that *in every case* the average **max** figure elicited by the SG procedure is strictly lower than the corresponding **min** figure elicited via CV. So here too we observe massive discrepancies between two sets of measures which share exactly the same theoretical foundations, and which therefore ought, according to standard economic theory, to be approximately equal.

The follow-up stage of the Main Study, where 101 respondents were visited again and presented with questions of whichever type they had not answered during the first visit, showed the disparities to be robust on a within-subject basis. And subsequent smaller scale studies with convenience samples[19] demonstrated that the disparities were easy to replicate even when respondents were more familiar with the prognoses and the elicitation techniques, and were above-averagely numerate.

However, although we now had considerable evidence that the CV-based estimates were heavily biased in an upward direction, and although we could not detect any such obvious biases in the SG responses, it was by no means certain that the SG-based figures really did represent people's preferences; nor were we able to do more than speculate about the reasons why people reacted so systematically differently to the two response modes.

In order to 'get under the skin' of what was happening, we embarked on a further series of studies[20] in which we not only asked people to answer via at least two response modes, but also subsequently interviewed a subset of those respondents to discover more about the processes they were using to arrive at their responses. This work is ongoing, so that some of the results are only in Working Paper/draft form, while others are not yet written up to even that extent. However, the next few paragraphs summarise some of the main points which have emerged so far.

Probably the single biggest reason for the disparities between CV- and SG-based estimates appears to be a form of reference point effect. When answering the CV questions, respondents tended to focus on the more severe features of the prognosis from the standpoint of their present state of health, and tended to think about the implications in terms of the losses they would sustain relative to their *status quo*. They tended *not* to think about how they would adjust - e.g. in terms of the new activities/sources of utility that they would substitute for the things they would lose. They also tended to be very myopic, in that they often focussed on the early (and worst) features of the prognosis and tended to pay little or no attention to the later phases where things might improve - even in the cases where they would return to normal health with no residual effects after the initial period of 1-3 years (in the case of X) or 3-4 months (for W).

By contrast, the SG form of question seemed to have the effect of switching their standpoint to one where they put themselves in the position of having already sustained the injury and now faced the prospect of a particular prognosis with certainty - unless they chose the risky alternative. Relative to this new reference point, the risky treatment offered the possibility of returning to normal health if the treatment succeeded; but it also involved the possibility of ending up in some even worse state if the treatment failed. According to subsequent interviews with a number of respondents, this had the effect of refocussing attention, sometimes quite dramatically. Not only did they give more weight to the things which they would still be able to do in the intermediate health state but which they stood to lose if the treatment failed, but they also tended to become much more conscious of the relative time spent in the different phases of each prognosis. For a 40-year-old faced with the prospect of injury W, this might well involve putting the expectation of a further 30-40 years of normal health in jeopardy if the treatment failed; whereas if they stuck to the certainty of W they would, without any risk at all, be back to normal health in just a few months.

The reference point/focus of attention switch was probably the most important single factor explaining the CV-SG disparities. But in a number of cases there were other effects, too. For example, a further factor that seems to have played a role may be called the "minimum credible response effect". In the case of the CV scenarios, it seemed that in the absence of sufficiently accessible and well-established pre-existing preferences, a number of respondents looked for cues about what would constitute a "reasonable" or "sensible" amount to pay: in the subsequent interviews, some mentioned specific safety devices that they had thought about - e.g. seat belts, cycle helmets, etc. And even if they had no particular device in mind, some expressed the view that "you couldn't get anything worth having for less than x", where x was unlikely to be less than £5 or £10, and might be a good deal more[21].

By contrast, the SG respondents had no inhibitions about saying that they would not take any risk of failure at all, particularly in the case of the temporary injuries, X and W: indeed, in the Main Study, zero risk of failure was the median response in both of these cases[22].

There were several other factors which may also have had some impact[23]. However, what has been described above is probably sufficient to make the main point. The fact that two response modes are rooted in the same theoretical foundations and ought to produce similar estimates is not enough: the theoretical equivalence is not apparent even to respondents with higher degrees in economics who are presented with both types of questions in the same questionnaire. So it is hardly surprising that members of the public at large who only see one of the two types of question, and who are attempting to provide acceptable responses to a series of unfamiliar and somewhat abstract hypothetical questions, may be strongly and systematically influenced by a variety of 'irrelevant' (but salient) features of the particular type of question and/or may resort to various imperfect but not obviously unreasonable rules of thumb to produce the necessary responses.

Additional Considerations

All of the discussion in the preceding sections has focussed on the elicitation of individuals' values for reductions in their *own* risks of death or injury in *road* traffic accidents. Two obvious questions arise. How (if at all) should values take account of the *distribution* of risks between different members of the population? And how (if at all) should different values be attached to reductions in risks to life and health in *other contexts*?

Let us begin with some distributional issues. Loomes (1982) and Keller and Sarin (1988) provide evidence that many people have preferences over different distributions of risk of death even when the expected number of fatalities and the respondent's own personal risk are held constant. For example, Loomes (1982) asked respondents to rank three scenarios, A, B and C: scenario A involved the certainty that 1 randomly chosen individual would die, with the other n-1 individuals sure to survive; B gave each of the n individuals an independent 1/n chance of dying; and C entailed a 1/n chance that all n individuals would die, with an (n-1)/n chance that they would all survive. Very few respondents expressed indifference between even two of these three scenarios; but neither was there a consensus about the ranking. What appeared to be happening was that those who placed a greater weight (or happened to focus) on *ex post equity*[24] tended to rank C above B above A, whereas those whose decision weight (or focus of attention) was placed upon damage limitation tended towards the opposite ranking - A over B over C.

In the example just cited, respondents were assigned to one of two subsamples. In one case, they were asked to imagine that the other n-1 individuals in the population at risk were people they had never previously met, whereas in the other subsample, respondents were asked to imagine that the n-1 others included their immediate relatives. This distinguishing factor also made a difference to the pattern of rankings. In addition to which, Keller and Sarin's (1988) study suggests that the *distribution of benefits in conjunction with the distribution of risks* also enters into people's preferences.

Besides such empirical work, there has also been a good deal of theoretical consideration of such issues - see, for example, Harvey (1985) or Fishburn and Straffin (1989), among others. In particular, the issue of whether and under what circumstances altruistic concerns ought or ought not to enter into the value of safety has recently been addressed by Jones-Lee (1991, 1992), who has identified various types of altruism and has suggested that "on the basis of not entirely implausible assumptions concerning the properties of the distribution of altruistic concern across the U.K. population, it transpires that one can, with some confidence, conclude that the value of statistical life for a 'caring' society will be some 10% to 40% larger than the value that would be appropriate for a society of purely self-interested individuals." (Jones-Lee, 1992, p.89)

Yet as we discovered during pre-piloting for the DoT study, it is one thing to identify the potential theoretical importance of altruistic concerns, and rather a different matter to devise questions which successfully tap into and measure such concerns. Indeed, we made so little progress with this task that (as outlined in previous sections) we opted instead for forms of question which explicitly asked people to disregard such considerations and focus exclusively upon their own safety. However, it is not clear how far some respondents were either willing or able to comply with this request - especially in the case of the CV questions[25]. Thus, notwithstanding attempts such as those by Viscusi et al (1988) to take account of altruistic concerns in CV-based values of safety, there is much work still to be done to examine the nature of such preferences and to devise appropriate ways of measuring the values people place on others' safety and incorporate the relevant values into the figures used for public policy purposes.

However, as noted earlier, besides distributional/altruistic issues there may also be a number of other contextual factors that need to be considered. Our starting point here is research on the factors which influence people's perceptions of risk and the degree to which different risks are felt to be (un)acceptable. An important stream of such work has been conducted within the kind of 'psychometric paradigm' employed by Slovic, Fischhoff and Lichtenstein (1980). In the course of such work, a number of factors have been identified as potentially influential in determining public perceptions of risk: for example, the extent to which particular risks are voluntary or involuntary, avoidable or unavoidable, natural or manmade, known or unknown, who is responsible for them, how far they are controllable by oneself or by others (or not controllable by anyone), and so on[26].

Such factors clearly enter into people's *perceptions* of risks. But how far do they enter into their *preferences* in a way that should (if it could) be measured and incorporated into public policy values and resource allocation decisions? For example, road transport risks may be widely regarded as largely voluntary, within the individual's area of personal responsibility and control, and with low potential for 'catastrophe'[27]. By contrast, air travel involves putting oneself much more obviously in other people's hands, and has greater catastrophic potential. Does it therefore follow that a higher value should be placed on a 'statistical life' in the context of air travel than in the context of road transport? And how might this compare with, say, the values used in the context of food safety? And within that context, might there be a case for valuing reductions in the risks of (natural) bacterial contamination differently from statistically comparable risks due to (manmade) pesticides? And should the (putative) risks from genetic engineering - e.g. putting scorpion venom genes into cabbages to paralyse marauding caterpillars - be valued differently again?

In practice there may be very substantial differences between the value of a statistical life implicit in different decisions made in a variety of different contexts. Often such disparities are unintentional - we may even be unaware of many of them - since they come about in the course of separate decisions made according to 'local' criteria, and/or precipitated by short-term political expediency, media scares, etc. Few would try to defend differences that arise in this way.

On the other hand, there is no reason *in principle* why preference-based values for preventing premature death (or injury, or illness) should be required to be the same, irrespective of context. If it is legitimate to trade off probability of death against other sources of welfare (e.g. time taken to complete a journey), then it can be argued that it is legitimate to trade off probability of death against some other characteristic of a risky situation, such as degree of control over the risk. Indeed, this point is made explicitly in the U.K. Treasury's (1991) Technical Guide[28].

However, having accepted the principle, the practical question remains as to how far such characteristics can be successfully incorporated into a single measure of value. We have seen that even when focussing on a single context - road safety - and asking respondents to think exclusively about their own risk, numerous serious difficulties arose. How then could one hope to extend the scope of the exercise to embrace considerations of equity, altruism, and all relevant contextual considerations? In the last few years, this question has started to receive increasing attention.

Recent and Future Developments

One way of responding to the various difficulties that have been encountered in many value elicitation studies is typified by the report of the NOAA Panel. While acknowledging - indeed, cataloguing - a number of the well-known problems and criticisms, the basic message of this report seems to be that these difficulties can be substantially overcome by improving survey design. Hence the Panel "try to lay down a fairly complete set of guidelines, compliance with which would define an ideal CV survey." (Arrow et al, 1993, Section IV)

In effect, then, the Panel are subscribing to a model of individual repondents somewhat similar to the standard (and rather idealised) portrait sketched out at the beginning of this paper: namely, that people actually do have 'true' preferences, and that although they may not necessarily be fully matured in every respect, sufficiently robust and unbiased answers can be obtained by means of a mass survey conducted according to the guidelines and standards set out by the Panel. This is an attitude to the public neatly (if ironically) summed up by Fischhoff (1991) with the maxim: "If we've got questions, then they've got answers." More formally, he refers to it as the *philosophy of articulated values.*

He contrasts this with what he calls the *philosophy of basic values*, which "holds that people lack well-differentiated values for all but the most familiar of evaluation questions ... In other cases, they must derive specific valuations from some basic values through an inferential process." (Fischhoff, 1991, p.835)

The philosophy of basic values is closely related to the notion of the *constructive* nature of preferences/decisions/responses - that is, the notion that rather than simply retrieving a well-formed pre-existing preference in response to a question or problem, individuals must to some extent engage in a *mental production process* in order to *create* their response. What is more, such responses may not always be generated by some invariant and widely-shared algorithm: rather, the particular algorithm used on any occasion may to some extent be influenced by the form and framing of the question, the (unintentional) cues provided, and the particular features of the problem that catch the respondent's attention at the time. Moreover, to the extent that individuals do sometimes call on particular ways of thinking about certain kinds of problems, these may often involve simplified rules of thumb or heuristics which may well be prone to systematic biases. Finally, if different heuristics are applied to different forms of questions, they may well generate patterns of behaviour which, when viewed as a whole, may fail to conform with any well-defined set of axioms or principles of regularity.

In the context of certain environmental scenarios, Schkade and Payne (1994) have reported evidence of the rich variety of considerations and cognitive strategies respondents employed to generate monetary value responses, as well as collecting data about the extent to which respondents found the questions difficult to answer and how confident they felt about their responses. As indicated earlier[29], a similar diversity of considerations appeared to enter into our samples' responses to the various types of question used to elicit values for health and safety.

In the face of this, how might we proceed? One approach is to start from the premise that preferences/responses *are* to some degree constructed, and then to try to build that construction process into the elicitation procedure in a more controlled way. Thus both Gregory et al (1993) and Baron (1994) advocate the use of decision analysis, in the form of multiattribute utility theory (MAUT) applied fairly intensively to produce a money value for the good or benefit in question. Fuller details are given in both papers, but broadly speaking this would entail: identifying the key attributes or dimensions of the good which tap into respondents' basic values; estimating utilities along each basic dimension; exploring the relative weights attached to these dimensions and the appropriate basis for combining them into a single utility score which can then be converted into a monetary value; and throughout, running various consistency checks - identifying discrepancies, uncovering their sources and trying to find some acceptable resolution.

Of course, as both papers acknowledge, such a procedure requires a great deal more time per respondent, so that sample sizes will necessarily be much smaller than those envisaged by the NOAA Panel and many current practitioners of contingent valuation. Gregory et al (1993, p.189) talk in terms of "... fewer than 100 people. Depth of value analysis is substituted for breadth of population sampling." No doubt, some researchers and policymakers will be nervous about such a relatively small sample combined with such a large input from the researcher/interviewer. It is certainly something which requires careful monitoring. On the other hand, as the early sections of this paper, and studies by many others, have clearly demonstrated, large random sample surveys by themselves are of little use if the methods they employ are vulnerable to substantial biases and inconsistencies which are poorly understood and difficult to adjust for.

Our own view is that we should certainly explore the route of smaller samples/panels that can be questioned more intensively, and see how far this approach can be developed and what its own limitations appear to be. Ideally, such studies should be conducted *in conjunction with* more conventional survey techniques, so as to compare the results and investigate the relationships between them.

However, even though the more intensive constructive approach may well yield many useful insights, we anticipate that it too will encounter substantial difficulties. Confining attention to the field of health and safety[30], let us briefly consider several aspects that seem potentially most troublesome.

It may seem that asking people about the values they place on personal health and safety should present fewer problems than asking them about the value of (often unfamiliar and somewhat remote) environmental goods. After all, health and safety are a vital and ever-present part of people's daily lives, and we may therefore expect them to have greater experience to call on, and a greater incentive to think seriously about such matters.

On the other hand, there are three key elements in health and safety, all of which might present serious difficulties which may not be easy to overcome even under the conditions of intensive interviewing entailed by the constructive approach.

First, there are the magnitudes of the risks. In most cases where they are known, they are very small - often expressed in terms of x chances in a million. By these standards, the road transport risks we presented to respondents were really quite large: and yet it was clear that many people had difficulty getting a genuine 'feel' for information such as "a baseline of 16 in 100,000 reduced by 4 in 100,000 down to 12 in 100,000". But what can or should be done about this? For example, should the information about the baseline and the residual risk be omitted in order to focus exclusively on the reduction of 4 in 100,000? We know that in one-off interviews many people appeared insensitive to the difference between 4 in 100,000 and 12 in 100,000 - but would thinking longer and harder about it really help? And what does it mean to say that this risk is 'known'? Really, what is being presented to people is a measure of the frequency of road accident fatalities in the population - but what people are then required to do is to somehow translate that information into terms relevant to their individual prospects, and we do not know much about that process of translation, and how far it may be influenced by the way(s) in which the information is presented.

And what about the many cases where the probabilities are not known - or at least, not easily quantifiable on any 'objective' basis? In many - perhaps nearly all - cases, the 'chances' of some event occurring are largely a matter of judgement, about which there is frequently disagreement. Will people's attitudes to such disagreements and ambiguities be allowable elements in the measurement of their preferences - in which case, how will they be incorporated? Or will they be identified as internal inconsistencies which will be 'resolved' during the intensive process of checking and reconciliation? Two questions (at least) arise here. First, who decides how many and which different judgements are presented to respondents? And second, how do we judge which patterns of response are taken to be expressions of basic values and which are taken to be in need of further refinement?

The more general and fundamental point we wish to make is that the skilful application of decision analysis may well reduce disparities and produce convergence - but convergence does not necessarily reveal 'truth': it may simply reflect the fact that when presented with apparent discrepancies between partially articulated responses, people may be persuaded that such discrepancies ought to be removed - and may therefore employ some additional algorithm for removing them. For example, suppose that (in classic preference reversal tradition) someone ranks A above B but places a higher money value on B than on A. When this is drawn to their attention, they may well feel that they ought to resolve this apparent anomaly; and in order to do so, perhaps they bring their value of B down below the value placed on A. But have they now revealed their *true* values? If both of the original values placed on A and B, and the original ranking, were to some degree constructed responses, to which they have now applied a constructed method of adjustment, what is the status of the final values that they have arrived at? True? Good enough? Or requiring further investigation?

Not (yet[31]) having an answer to that question, let us turn instead to the second key element of many health and safety decisions: the valuation of health states. As the surveys cited in footnote 4 show, numerous studies have used a variety of techniques to produce values/utilities for a large range of health states inferior to normal health. This literature is rife with disparities and anomalies, suggesting that many respondents cannot readily attach values to the quality of life entailed by states of health unfamiliar to them. And indeed, it is a daunting task: besides identifying those activities which one does now but would not be able to do (at least, not to the same extent) in some other health state, one ideally needs to anticipate what new activities would be substituted for the old and how much utility would be derived from them. This latter task in particular makes heavy demands on respondents' powers of imagination and anticipation, and there is reason to be cautious about the validity of what it may produce. In a section called 'Predicted Utility - Do People Know What They Will Like?', Kahneman (1994) cites a number of studies, including some involving health states, which suggest that the answer to that question is very often 'No': and if this seems to be true for even quite narrowly focused and relatively familiar cases such as the consumption of ice cream and yogurt, how much more circumspect should we be about people's abilities to anticipate the utility of living in different health states which may require drastic and fundamental reorganisation of their lives in many important respects? It will be interesting - and important - to find out.

The third element - which cannot easily be disentangled from the second (and which will therefore pose problems for procedures which try first to decompose and then reassemble the different elements) is time. As the work of Loewenstein and Prelec (1991, 1992) and others[32] has shown, there seem to be just as many anomalies in intertemporal choice as there are in choice under uncertainty. In the context of health, Bleichrodt and Gafni (1994) argue persuasively that standard ways of modelling time preference are neither normatively appealing nor descriptively accurate. If MAUT is applied in a way which tries to force time preferences into some standard mould, it may well run into grave difficulties. On the other hand, if - as we hope and expect - it is used to gain deeper insights into what is at present a contentious, even mysterious, domain, it may enlighten and inform theory and policy in very significant ways.

Concluding Remarks

Although it is clearly desirable to value the benefits of reducing risks to health and safety in some way which allows those benefits to be given due weight in decisions about the allocation of scarce resources, we have seen that the task of generating such values presents a substantial number of theoretical and practical challenges.

The evidence cited above (and much more besides) suggests that in these matters it is certainly not safe to proceed as if most members of the population have stable, well-behaved and highly articulated preferences which can be readily accessed by standard questionnaire surveys. While we should not necessarily abandon the questionnaire approach (and indeed, in the short run at least, we should make every attempt to improve the quality of such surveys), we must also be prepared to explore some more radical alternatives and address some more fundamental issues.

For example, we need to ask at what level - *if at all* - individuals really do have something like stable and well-behaved preferences. How basic do we have to get? How precise can respondents be - and how do we accommodate their imprecision? And if it turns out that, with sufficient opportunity and incentive to consider their responses carefully, many people *do* exhibit preferences which tend to conform with some principles of regularity, how far are these consistent with sets of axioms such as those of von Neumann and Morgenstern, or how far do they correspond with one or other of the many non-expected utility models? If the answer to this last question is either that most people do not converge towards some stable set of preferences, or else that most people do tend towards responses consistent with one or other well-defined model but that different individuals conform with different (often non-expected utility) models, how should we handle this result? Such questions will, we believe, stimulate much important and interesting theoretical and practical research during the next decade.

Acknowledgements

This paper draws on work done with a number of collaborators - in particular, Paul Dolan, Richard Dubourg, Deirdre O'Reilly, Peter Philips and Angela Robinson. Support for various parts of the work came from the Department of Transport and the Economic and Social Research Council, Awards R000231194 and R000234987. The opinions and conclusions contained in this paper are those of the authors, and do not necessarily reflect the views of DoT or ESRC.

Footnotes

1. In Annex B of H.M.Treasury's (1991) Technical Guide for Government Departments, the valuation of the avoidance of premature death is given as an example of a case where "confounding factors combined with the possibility of misperception by consumers of the probabilities of premature death make estimations of the revealed value of life very uncertain." The next paragraph goes on to say: "A direct questionnaire approach can, by careful design, avoid many of the confounding factors ..." (Paragraphs 28 & 29). The U.S. NOAA Panel on Contingent Valuation - see Arrow et al (1993) - have also recently endorsed such studies, setting out guidelines for what they consider to be 'good practice' in this respect. More will be said about some of these recommendations in due course.

2. Suppose that the sample mean value for a 4 in 100,000 reduction in the risk of death were £80: dividing the mean value by the size of the risk reduction gives a "value for preventing a 'statistical' death" of £2 million.

3. For a review of the considerations behind this change of policy, see Dalvi (1988). Although other evidence was also taken into account, a major influence on the Department of Transport's new value for preventing road accident fatalities was a national sample survey it commissioned in 1982, the essential features of which are reported in Jones-Lee et al (1985).

4. For surveys of the various techniques commonly used in health state measurement exercises, see, for example, Froberg and Kane (1989) and Brooks (1991).

5. We used this form of words to elicit an estimate of the point of indifference rather than a phrase such as "the maximum amount you would be just prepared to pay" since we felt that it better conveyed the notion of indifference to a lay audience, and was less likely to confuse respondents who were already being asked about the most they were sure they would pay, in order to establish their **min** value.

6. Two other forms of question - Standard Gamble and Risk-Risk - were also tried out in the first phase of piloting. Standard Gambles were also used in the subsequent Main Study, and will be described more fully in a later section of this paper. A description of

the Risk-Risk questions can be found in Dolan et al (1995), together with an account of the difficulties encountered, which were so severe that this type of question was not used again throughout the rest of the study.

7. The 1982 study of fatalities (again, see Jones-Lee et al, 1985) used a lump-sum scenario, and in order to maintain a link with that study, we also used such a scenario as part of the first phase of piloting. However, it entails various complications - not least, that respondents are often uncertain how long they expect to keep their present vehicle and therefore how long they expect to derive the benefits of the safety feature. Comparisons between the lump-sum and the annual payment scenarios showed that respondents were largely ignoring this time dimension (or else that they were discounting their futures at a phenomenally high rate). There were other possible problems, too: for example, respondents who drove company (fleet) vehicles might well demur at being asked to pay for a feature that would enhance the value of a vehicle they didn't own, while those who owned their own vehicles might include in their values the expected increase in resale price of the vehicles; and so on. The annual payment for an annual reduction, although somewhat artificial, was 'cleaner' and seemed no less acceptable to most respondents.

8. For earlier examples of starting point effects, see Rowe et al (1980) or Boyle et al (1985); by contrast, a study which found no significant starting point bias is reported by O'Brien and Viramontes (1992).

9. More will be said later about the idea of responses being constructions rather than revelations of preferences. But for now, this footnote will be confined to a more specific point relating to the report of the NOAA Panel on Contingent Valuation. One of the main recommendations of that report was that a good study design should involve a dichotomous-choice referendum format, with (very) large samples randomised between subsamples, each of which is presented with a question in the form of "would you be prepared to vote for a tax of x per head?" to provide some public good, where x is varied from one subsample to another over a range which in theory allows the underlying demand function to be estimated. However, strong starting point effects of the kind that we and others have observed suggest that there may be a serious overdisposition to say "yes" to such a question, so that the inferred demand function may owe much more to the set of values of x chosen by the study designers than to the 'true' preferences of the respondents being sampled. On these grounds alone, the Panel's apparently rather uninhibited advocacy of the dichotomous-choice referendum format seems, to put it mildly, somewhat premature. And as Baron (1994) has pointed out, there may also be other reasons for being more cautious about such a recommendation.

10. See Dubourg, Jones-Lee and Loomes (1994) for further details.

11. Some recent experimental data reported by Shogren et al (1994) has been interpreted by them as supporting Hanemann's analysis. However, their data are no less consistent with the kind of account outlined in the previous paragraph.

12. Again, see Dubourg, Jones-Lee and Loomes (1994) for details.

13. In the first phase of piloting, both of the direct valuation questionnaires used the same set of risk reduction scenarios as was presented to Groups A & B - and much the same pattern was observed, with $m_W/m_K > m_X/m_K$ for 55 out of the 63 individuals whose responses were usable for these computations.

14. It may be suspected that the bulk of these were zero 'protest' responses, but that is not so: only one respondent gave a string of zero responses.

15. That is, they were asked to consider the six showcards displaying the six scenarios, and set them out on a table, putting the one they would most prefer to have if they could only have one of the six at the top, and so on down to the one that they would have the weakest preference for at the bottom (with ties allowed). In other words, at this stage money values did not come into it - it was entirely a matter of balancing magnitudes of risk reduction against severities of prognoses.

16. This is, of course, a form of what Kahneman and Knetsch (1992) refer to as the embedding effect, where respondents state that they are willing to pay as much for some benefit X as for some unambiguously greater benefit Y which contains X as a component.

17. See Jones-Lee et al (1993) for a fuller report of the Main Study; and see Jones-Lee et al (1995) for additional analysis and commentary.

18. In some questions, the consequence of failure of the risky treatment was another non-fatal prognosis rated by most respondents as worse than the 'certain' prognosis.

19. These included classes of graduate students studying for a Masters degree in Health Economics, and a group of professional economists working in various U.K. Government Departments.

20. Funded by the Economic and Social Research Council, Awards R00023 4305 and R00023 4987. The End of Award Report relating to the first of these awards - Jones-Lee et al (1994) - is available on request.

21. Similar tendencies to look for clues about what might constitute a reasonable response, including considering the cost of producing a good rather than its value to the individual respondent, have been reported to be quite common in other contexts - see, for example, Schkade and Payne (1994) and Baron (1994),

22. Note that these were not for the most part 'protest' responses: the great majority of respondents who gave zero responses with respect to injuries X and W also registered a willingness to take some risk of failure in questions where the certain prognosis was

sufficiently severe and/or when the consequences of failure were more severe but not fatal.

23. For further details, see Jones-Lee et al (1994, p.8).

24. That is, how equitable is the distribution of harm after the uncertainty has been resolved.

25. Some respondents expressed a distaste for the idea of spending (household) money on a device that would give them some protection but offer no benefit to other people in the vehicle.

26. For helpful overviews, see Pidgeon et al (1992) or Slovic (1992).

27. Usually understood to mean an event where a number of people - informally, ten or more - are killed at the same time.

28. Paragraph 32 of Annex B reads: "The values for the avoidance of premature death and the cost of injuries in road safety applications used by the Department of Transport may not be directly applicable to other programmes. Individuals are not likely to be indifferent about the causes and circumstances of premature death and will compare, say, sudden death with death following painful illness. Different groups may also have different attitudes to risk, ... or may be influenced by the degree of control they feel they or others have over their own safety. However, if tailored to these circumstances, the general approach may have wider applications."

29. And set out in more detail in Jones-Lee et al (1994).

30. Of course, on past and present form it seems likely that the bulk of such studies in the near future may focus on environmental issues, where there may be somewhat different concerns. However, we should not be surprised to see some shift in the weight of research towards health and safety in due course.

31. Jane Beattie, Robin Cubitt, Graham Loomes, Chris Starmer and Robert Sugden have ESRC funding to investigate such questions under Award L122 25 1024: The Consistency or Inconsistency of Preferences Under Risk and Over Time.

32. The book edited by Loewenstein and Elster (1992) contains a number of other stimulating contributions in this area.

References

Arrow, K., Solow, R., Portney, P., Leamer, E., Radner, R. and Schuman, H., (1993), Report of the NOAA Panel on Contingent Valuation, *Federal Register*, **58**, 4602-14.

Baron, J., (1994), Rationality and Invariance: Response to Schuman, paper presented to DoE/EPA Workshop on Contingent Valuation, mimeo.

Bleichrodt, H. and Gafni, A., (1994), Time Preference, the Discounted Utility Model and Health, mimeo.

Boyle, K.J., Bishop, R.C. and Welsh, M.P., (1985), Starting Point Bias in Contingent Valuation Bidding Games, *Land Economics*, **61**, 188-94.

Brooks, R.G., (1991), Health Status and Quality of Life Measurement: Issues and Developments, Lund:Swedish Institute of Health Economics.

Brookshire, D.S. and Coursey, D.L., (1987), Measuring the Value of a Public Good: An Empirical Comparison of Elicitation Procedures, *American Economic Review*, **77**, 554-66.

Coursey, D.L., Hovis, J.L. and Schulze, W.D., (1987), The Disparity Between Willingness To Accept and Willingness To Pay Measures of Value, *Quarterly Journal of Economics*, **102**, 679-90.

Dalvi, M.Q., (1988), The Value of Life and Safety: A Search for a Consensus Estimate, London:Department of Transport.

Dolan, P., Jones-Lee, M.W. and Loomes, G., (1995), Risk-Risk vs Standard Gamble Procedures for Measuring Health State Utilities, *Applied Economics*, **27**, 1103-11.

Dubourg, W.R., Jones-Lee, M.W. and Loomes, G., (1994), Imprecise Preferences and the WTP-WTA Disparity, *Journal of Risk and Uncertainty*, **9**, 115-33.

Dubourg, W.R., Jones-Lee, M.W. and Loomes, G., (1995), Imprecise Preferences and Survey Design in Contingent Valuation, mimeo.

Fischhoff, B., (1991), Value Elicitation: Is There Anything In There?, *American Psychologist*, **46**, 835-47.

Fishburn, P.C. and Straffin, P.C., (1989), Equity Considerations in Public Risk Evaluation, *Operations Research*, **37**, 229-39.

Froberg, D.G. and Kane, R.L., (1989), Methodology for Measuring Health State Preferences, I-IV, *Journal of Clinical Epidemiology*, **42**, 345-54, 459-71, 585-92, 675-85.

Gregory, R., Lichtenstein, S. and Slovic, P., (1993), Valuing Environmental Resources: A Constructive Approach, *Journal of Risk and Uncertainty*, **7**, 177-97.

Hanemann, W.M., (1991), Willingness To Pay and Willingness To Accept: How Much Can They Differ?, *American Economic Review*, **81**, 635-47.

Harvey, C., (1985), Decision Analysis Models for Social Attitudes Toward Inequity, *Management Science*, **21**, 1199-1212.

H.M.Treasury, (1991), Economic Appraisal in Central Government: A Technical Guide for Government Departments, London:HMSO.

Jones-Lee, M.W., (1991), Altruism and the Value of Other People's Safety, *Journal of Risk and Uncertainty*, **4**, 213-9.

Jones-Lee, M.W., (1992), Paternalistic Altruism and the Value of Statistical Life, *Economic Journal*, **102**, 80-90.

Jones-Lee, M.W., Hammerton, M. and Philips, P.R., (1985), The Value of Safety: Results of a National Sample Survey, *Economic Journal*, **95**, 49-72.

Jones-Lee, M.W., Loomes, G., O'Reilly, D. and Philips, P.R., (1993), The Value of Preventing Non-Fatal Road Injuries: Findings of a Willingness-To-Pay National Sample Survey, *Transport Research Laboratory Contractor Report* **330**, London:HMSO.

Jones-Lee, M.W., Loomes, G., and Philips, P.R., (1995), Valuing the Prevention of Non-Fatal Road Injuries: Contingent Valuation vs Standard Gambles, *Oxford Economic Papers*, **47**, 676-95.

Jones-Lee, M.W., Loomes, G., and Robinson, A., (1994), The Relationship Between Different Methods of Estimating the Value of Safety, *ESRC End of Award Report*, mimeo.

Kahneman, D., (1994), New Challenges to the Rationality Assumption, *Journal of Institutional and Theoretical Economics*, in press.

Kahneman, D., Knetsch, J.L. and Thaler, R.H., (1991), The Endowment Effect, Loss Aversion, and Status Quo Bias, *Journal of Economic Perspectives*, **5**, 193-206.

Kahneman, D. and Knetsch, J.L., (1992), Valuing Public Goods: The Purchase of Moral Satisfaction, *Journal of Environmental Economics and Management*, **22**, 57-70.

Keller, L.R. and Sarin, R.K., (1988), Equity in Social Risk: Some Empirical Observations, *Risk Analysis*, **8**, 135-46.

Knetsch, J.L. and Sinden, J.A., (1984), Willingness To Pay and Compensation Demanded: Experimental Evidence of an Unexpected Disparity in Measures of Value, *Quarterly Journal of Economics*, **99**, 507-21.

Loewenstein, G. and Elster, J., (1992), Choice Over Time, New York:Russell Sage.

Loewenstein, G. and Prelec, D., (1991), Negative Time Preference, *American Economic Review Proceedings*, **81**, 347-52.

Loewenstein, G. and Prelec, D., (1992), Anomalies in Intertemporal Choice: Evidence and an Interpretation, *Quarterly Journal of Economics*, **107**, 573-97.

Loomes, G., (1982), Choices Involving a Risk of Death: An Empirical Study, *Scottish Journal of Political Economy*, **29**, 272-82.

McDaniels, T.L., (1992), Reference Points, Loss Aversion and Contingent Values for Auto Safety, *Journal of Risk and Uncertainty*, **5**, 187-200.

Marin, A., (1992), Appendix: Cost and Benefits of Risk Reductions, in *Risk: Analysis, Perception and Management*, 192-201, London:The Royal Society.

O'Brien, B. and Viramontes, J.L., (1992), Willingness-To-Pay: A Valid and Reliable Measure of Health State Preference?, mimeo.

Pidgeon, N., Hood, C., Jones, D., Turner, B. and Gibson, R., (1992), Risk Perception, in *Risk: Analysis, Perception and Management*, 89-134, London:The Royal Society.

Rowe, R.D., d'Arge, R.C. and Brookshire, D.S., (1980), An Experiment on the Economic Value of Visibility, *Journal of Environmental Economics and Management*, **7**, 1-19.

Schkade, D.A. and Payne, J.W., (1994), How People Respond to Contingent Valuation Questions: A Verbal Protocol Analysis of Willingness To Pay for an Environmental Regulation, *Journal of Environmental Economics and Management*, **26**, 88-109.

Shogren, J.F., Shin, S.Y., Hayes, D.J. and Kliebenstein, J.B., (1994), Resolving Differences in Willingness To Pay and Willingness To Accept, *American Economic Review*, **84**, 255-70.

Slovic, P., (1992), Perception of Risk: Reflections on the Psychometric Paradigm, Ch. 5 in *Social Theories of Risk*, eds. Krimsky, S. and Golding, D., Westport:Praeger.

Slovic, P., Fischhoff, B. and Lichtenstein, S., (1980), Facts and Fears: Understanding Perceived Risk, in *Societal Risk Assessment: How Safe is Safe Enough?*, eds. Schwing, R. and Albers, W., New York:Plenum Press.

Thaler, R., (1980), Toward a Positive Theory of Consumer Choice, *Journal of Economic Behavior and Organization*, **1**, 39-60.

Tversky, A. and Kahneman, D., (1991), Loss Aversion in Riskless Choice: A Reference Dependent Model, *Quarterly Journal of Economics*, **107**, 1039-61.

Viscusi, W.K., Magat, W.A. and Huber, J., (1987), An Investigation of the Rationality of Consumer Valuations of Multiple Health Risks, *RAND Journal of Economics*, **18**, 465-79.

Viscusi, W.K., Magat, W.A. and Forrest, A., (1988), Altruistic and Private Valuations of Risk Reduction, *Journal of Policy Analysis and Management*, **7**, 227-45.

Appendix A: Injury Prognoses

R

In Hospital
- several weeks, possibly several months
- moderate to severe pain

After Hospital
- continuing pain/discomfort for the rest of your life, possibly requiring frequent medication
- substantial and permanent restrictions to your work and leisure activities--possibly some prominent scarring

S

In Hospital
- 1-4 weeks
- moderate to severe pain

After Hospital
- Some pain gradually reducing, but may recur when you take part in some activities
- some restrictions to leisure and possibly some work activities for the rest of your life

W

In Hospital
- 2-7 days
- slight to moderate pain

After Hospital
- some pain/discomfort for several weeks
- some restrictions to work and/or leisure activities for several weeks/months
- after 3-4 months, return to normal health with no permanent disability

X

In Hospital
- 1-4 weeks
- slight to moderate pain

After Hospital
- some pain/discomfort, gradually reducing
- some restrictions to work and leisure activities, steadily improving after 1-3 years, return to normal health with no permanent disability

IMPROVING EFFICIENCY IN ENVIRONMENTAL ENFORCEMENT

D. Marc KILGOUR
Wilfrid Laurier University
Waterloo, Ontario N2L 3C5 Canada

Abstract: A livable environment can be achieved, and preserved, only if new and better methods of regulating individual choices are developed to balance private and social goals. Game-theoretic models show how the effectiveness of regulations, and their costs, are determined by the strategic interaction of the enforcement agency with the subjects of the regulation. These models are used to assess measures that could improve the efficiency of enforcement systems for environmental regulations.

1. Introduction

Humanity's greatest challenge may well be the crucial need to attain and maintain a livable environment. To meet this challenge, we will need to ensure that human choices take into account all of the consequential costs and benefits to current and future generations. We must develop and apply measures that balance short-term against long-term interests, private against public goods, and individual against social goals. To this end, it is inevitable that more and more regulations will constrain human decisions affecting the environment.

Most approaches to environmental regulation are of the traditional "command-and-control" type, in which some decision choices are explicitly "prohibited," using legal or quasi-legal means including the threat of sanctions. Alternative measures, such as transferable rights to discharge up to a specified amount of a regulated contaminant, are often attractive alternatives. But, no matter how actions affecting the environment are regulated, constraints must be imposed on human decision choices. Laws and regulations are crucial both to attaining and to maintaining a livable environment.

But simply developing new rules is not enough. Regulations "are only as good as their enforcement," as *Canada's Green Plan* (1990) put it. In other words, it is rarely easy to impose effective rules to prevent individuals from pursuing their personal interests to the fullest. A rule must be formalized and given official force, the decision makers whose choices it constrains must be informed, and *then they must choose to comply.*

Strictly speaking, of course, it is unlikely that 100% compliance to large-scale public regulations is ever achieved. The anticipated level of compliance is only one factor to be taken into account when a regulation is designed. Other important considerations include the exact contents (level and extent) of the regulation, the selection (or formation) of an

R. Nau et al. (eds.), Economic and Environmental Risk and Uncertainty, 33–52.

enforcement agency, the availability and reliability of compliance measurements, the procedures for judging compliance or noncompliance, and the sanctions for noncompliance.

The present work focuses on enforcement - how an enforcement agency should carry out its mandate of identifying instances of noncompliance to a regulation and applying legal or other procedures when an apparent violation has occurred. The primary objective of any enforcement agency, especially in the environmental field, must be to obtain as high a level of compliance as possible, thereby maximizing benefits to the environment.

But the *effectiveness* of enforcement, as measured by compliance level, is not the only criterion for success. An enforcement agency has a budget and other resources, so the *costs* of its efforts to enforce a regulation can be measured, at least in principle. Thus, whatever the level of compliance, a secondary objective of the agency is to minimize costs. It is useful, in fact, to combine effectiveness and cost into a single measure of success, *efficiency,* that reflects both. For instance, efficiency can be increased by making enforcement more effective (i.e. increasing compliance) without increasing costs, or by reducing costs without reducing compliance. If the efficiency of enforcement is maximized, then the level of enforcement resources necessary to achieve the objectives of a regulation is minimized, making more regulations feasible, freeing resources for other purposes, and making it more likely that environmental improvements will be achieved and protected.

Environmental regulations that cannot be enforced efficiently are of dubious value for other reasons. There is plenty of evidence that individuals with greater resources make choices that tend to result in longer and healthier lives. The costs of enforcing a regulation are ultimately drawn from individuals, who thus suffer a reduction in resources, which indirectly reduces health and lifespan. If the social benefits of a regulation, taking into account the level of compliance, are insufficient to compensate for this reduction, then the regulation is hard to justify on social grounds. (Keeney, 1994) Of course, there are many possible benefits of regulations that should be weighed into the balance, such as the rights of future generations, the need for sustainability, and the preservation of biodiversity.

In carrying out its enforcement obligations, an environmental agency typically monitors publicly available data, such as water quality in a stream, and then follows up with specific inspections of an operator's premises when the data warrants. (Russell *et al.* 1986, Fang *et al.* 1994) In general, environmental monitoring data is subject to a high degree of statistical and other variation, so that convincing conclusions are difficult to obtain in many instances. More accurate monitoring systems often have relatively high start-up (fixed) costs, although in some cases there are substantially lower direct (variable) costs. Some very cheap information gathering systems, such as waiting for a whistle-blower, are widely used and can be effective, even though there is considerable risk that the information gained is inaccurate, or even wrong. A cost-conscious agency commonly tries to draw as much useful information as possible from less costly monitoring data, avoiding the more expensive (and politically risky) premises inspections unless solid evidence of a violation is likely to result.

Two important but independent aspects of enforcement are the agency's procedures when it believes it has evidence of violation, and the sanctions that may be imposed on the operator as a result. There have been several studies of the consequences of punishment levels and uncertainties. (Kilgour *et al.* 1992) Some have argued that the threat of more severe sanctions can reduce other enforcement costs, but this view is controversial. For instance, Ehrlich (1973) found evidence in data on participation in illegal activities that greater penalties do result in lower crime rates. But more recent studies, such as Andreoni (1994), suggest that greater penalties also result in lower conviction probabilities, so their deterrent effects may be marginal. Interacting choices such as these can be represented formally only in a game-theoretic model, one of the first of which appeared in the provocatively-titled article, "Penalty has no impact on crime." (Tsebelis 1990)

The objective of this paper is to develop non-cooperative game models that address fundamental questions related to the enforcement of environmental regulations. The game-theoretic framework includes the essential strategic features of enforcement:

- the operator's comply/violate decision, based on a comparison of the benefits of violation versus its consequences, including the magnitude and likelihood of penalties and other costs,
- the agency's accept/inspect/accuse decisions, based on evidence that may contain errors (of either type), made subject to pressures to minimize costs, and
- the interaction of these decisions.

The game models here capture these strategic considerations in a very simple framework. They extend and elaborate Kilgour's (1994b) model for generic enforcement problems in the context of costless though imprecise monitoring and/or inspection data. A crucial difference is that here all information, however inaccurate, is costly.

2. The Basic Enforcement Model

Enforcement is modeled as a non-cooperative two-person game of complete information, with players *Subject* (S) and *Agency* (Ag). Their decisions concern Subject's choice to *Comply with* (C), or *Violate* (V), a regulation. Note that, in these models, a Subject who violates has no further choices, such as when, where, or how to violate. Thus, Subject is deprived of any actions that might help to conceal a violation, and low levels of violation, a possible cause of ambiguity, need not be considered.

The second player, Agency, also has a primary choice between two actions: it can either *Accept* (Ap) Subject's behavior, or *Accuse* (As) Subject of violating. (Agency's prior decision to inspect is modeled shortly.) Just as there is only one level and mode of violation in the Basic Model, there is only one kind of accusation - so Agency never has to decide what kind of charges to bring, or what level of penalty to ask for.

Because Subject has two choices, Comply or Violate, and Agency has two choices, Accept or Accuse, this simple conflict has four outcomes, which are tabulated below.

	Symbol	Subject	Agency	Utilities
Accepted Compliance	(C, Ap)	Comply	Accept	(0, 0)
False Accusation	(C, As)	Comply	Accuse	$(-F_S, -F_A)$
Successful Violation	(V, Ap)	Violate	Accept	$(T, -L)$
Apprehension	(V, As)	Violate	Accuse	$(-P, -M)$

Each player's preferences over the four outcomes are expressed by its (von Neumann-Morgenstern) utility values, shown (Subject's first) in the column on the right. The **Accepted Compliance** outcome is taken as the status quo; other utilities are measured relative to it.

We now explain the relationships utilities will be assumed to satisfy. Because false accusation would probably reflect negatively on Agency, the **False Accusation** outcome is less preferred by Agency relative to the status quo, or $F_A > 0$. Likewise, Subject will usually face defense costs and the risk of unfair punishment at this outcome, so its utility is less than at the status quo, implying $F_S > 0$. F_A and F_S are called the *False Accusation Costs.*

We assume here that subjects are motivated to violate the regulation - otherwise there would be no enforcement problem. Thus, Subject's utility for **Successful Violation,** the *Temptation, T,* must satisfy $T > 0$. Similarly, an agency suffers some costs when there has been a violation - reflecting bad publicity, clean-up, etc. - so Agency's *(Direct) Loss, L,* must satisfy $L > 0$.

Finally, when a violation has been apprehended, the violator usually faces a high likelihood of penalty as well as negative publicity, defense costs, clean-up costs, etc., outweighing the gain of violation. Thus *S*'s utility for **Apprehension** is typically negative, so *P*, the *Punishment,* is typically positive. On the other hand, the social loss consequent on a violation is reduced by its apprehension; as well Agency receives good publicity, though it may bear costs in imposing the penalty. In all, it can be assumed that Agency's *Mitigated Loss, M,* is positive but less than its Direct Loss, i.e. that $0 < M < L$.

Note that our assumptions imply that Agency is a perfect agent of society - it wants to prevent violations, and it wants to accuse in the event that a violation does occur. Thus the central issues of Agency Theory have been assumed away. It should be noted that one recent study suggests that these problems may be serious in the context of income tax, especially when convictions are hard to obtain. (Galantine 1994) We ignore these considerations here, primarily because in the environmental arena the usual result of an accusation is not a conviction decision, but a negotiation between the agency and the accused. (Hipel *et al.* 1994)

In summary, we have assumed that

$$\text{(Agency)} \qquad F_A > 0 \text{ and } L > M > 0.$$

$$\text{(Subject)} \qquad F_S > 0,\ T > 0, \text{ and } P > 0.$$

Thus, all payoff parameters are positive. Throughout we will ignore transitional cases; the consequences of, for example, equalities among the parameters, will not be considered.

As implied above, both players have full information about the game; in particular, they are aware of their opponent's, as well as their own, utilities. The only uncertainties are strategic - the players never learn each other's choices until the game is over (except as specified below) - or are the result of inaccurate inspections, which will be modeled below.

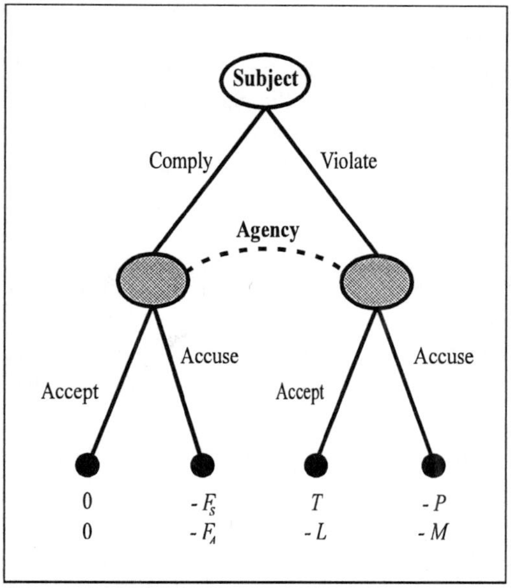

Figure 1: Enforcement without Inspection

An example used throughout for illustration can be introduced now. This example is a highly stylized environmental inspection problem, in which Agency is given a budget of, say, 100 units of utility, and is charged whenever there is a false accusation or a violation. Specifically, Agency's False Accusation Cost is $F_A = 1$. Its loss if there is a Successful Violation is $L = 9$, but its Mitigated Loss (if there is Apprehension) is only $M = 4$. Agency's utility is additive. Thus, Agency wishes to deter violations, and to apprehend any violations that do occur. It would tolerate some false accusations in order to deter violations. The players' inspection costs will be introduced below.

For Subject, as for Agency, the unit of utility is the False Accusation Cost, F_S. Temptation, T, and Penalty, P, are allowed to vary relative to F_S, in order to highlight the consequences of the most important initial choices facing the authority when it sets up the enforcement system. It is easy to understand the importance of Punishment, P, reflecting mainly the likelihood and the severity of penalty. As well, the relation of P with the Temptation, T, is often crucial. In many applications the authority implicitly chooses the temptation value, T, at least approximately. For example, if the gain from a violation depends on the length of time spent violating, and if violations are hard to conceal from an inspector, the primary determinant of T is the (mean) interval between inspections - which reflects the authority's choices of size, budget, and inspection rules for the agency.

Before introducing models, including inspection, it is useful to assess how enforcement would be carried out without benefit of inspection. This situation is Model A of Kilgour (1994b), shown as an extensive game in Figure 1. [Extensive games will be displayed in the notation of Fudenberg and Tirole (1991).] As explained in Kilgour (1994b), the game of Model A always has a unique (Nash) equilibrium. Let x be the probability that S chooses C and let y be the probability that Ag chooses Ap. If $P < F_S$, the equilibrium is $x = 0$, $y = 0$; if $P > F_S$, it is

$$x = \frac{L - M}{L - M + F_A} \; ; \; y = \frac{P - F_S}{P + T - F_S}$$

Thus, in Enforcement with No Possibility of Inspection there will be no compliance at all if the punishment value is too low, and partial compliance if the punishment value is high enough. One of two behavior patterns emerges, determined by the parameter values. The results for this approach in the Example case are shown in Figure 4.

3. Models of Inspection Systems

An inspection system is a procedure that provides Agency with a signal conditioned on Subject's choice of Comply or Violate. We consider here only binary inspection systems, which signal either Red or Green. A Red signal is an indication that Subject chose Violate, and a Green signal that Subject chose Comply. A perfect inspection system is one that always signals accurately:

$$Pr\{\text{Red}|\text{Violate}\} = 1, \qquad Pr\{\text{Red}|\text{Comply}\} = 0,$$

$$Pr\{\text{Green}|\text{Violate}\} = 0, \qquad Pr\{\text{Green}|\text{Comply}\} = 1.$$

Imperfect inspection systems, on the other hand, may result in errors. By analogy with classical statistics, there are two kinds of errors: Type I ("false alarm"), a Red signal when Subject chooses Comply; and Type II ("missed violation"), a Green signal when Subject chooses Violate. The error probabilities, α and β, describe the inspection system as follows:

$$Pr\{\text{Red}|\text{Violate}\} = 1 - \beta, \qquad Pr\{\text{Red}|\text{Comply}\} = \alpha,$$

$$Pr\{\text{Green}|\text{Violate}\} = \beta, \qquad Pr\{\text{Green}|\text{Comply}\} = 1 - \alpha.$$

Except for costs, any binary inspection system is completely characterized by its error probabilities, (α, β). A perfect inspection system has $(\alpha, \beta) = (0, 0)$. The smaller the values of α and β, the better. Not all error probability pairs need to be considered in practice. Suppose that α_r and β_r are non-negative numbers satisfying $\alpha_r + \beta_r = 1$. An inspection system with error probabilities (α_r, β_r) can be achieved by using a random procedure - signal Red with probability α_r, and Green with probability $\beta_r = 1 - \alpha_r$. Below, only "better than random" inspection systems, with $\alpha + \beta < 1$, will be considered, except that $(\alpha, \beta) = (0, 1)$ or $(1, 0)$ will be allowed. For such systems, (α, β) is restricted to the triangle shown in Figure 2.

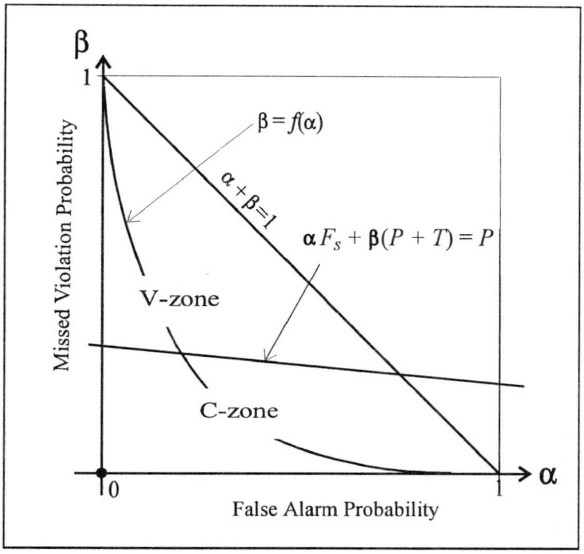

Figure 2: Inspection Error Probabilities

Many inspection systems are adjustable: they can be set to produce a Red signal even when the evidence of violation is weak, or only when it is very strong. Generally, a lower missed violation probability (β) is traded for a higher false alarm probability (α). Thus, many (α, β) pairs can be achieved using the same inspection system. Under plausible assumptions, that the best achievable (α, β) pairs are those with $\beta = f(\alpha)$, where the *characteristic* $f(\cdot)$ is a continuous, decreasing, weakly convex function defined for $\alpha \in [0, \alpha_0]$, satisfying $f(0) = \beta_0$ and $f(\alpha_0) = 0$. The intercepts α_0 and β_0 satisfy $0 < \alpha_0, \beta_0 \leq 1$. A typical characteristic is shown in Figure 2.

The other relevant characteristics of a binary inspection system are its costs, c_S to Subject and c_A to Agency. These represent the resources used by Agency in obtaining the inspection system signal, and the reduction in Subject's resources consequent on being inspected. Again, these payoff parameters are assumed non-negative.

When costs are taken into account, therefore, any inspection system can be characterized by its values for α, β, c_S, and c_A. Below, the example introduced earlier will be analyzed for a system with Type I error probability $\alpha = 0.3$, Type II error probability $\beta = 0.2$, and Agency's inspection cost $c_A = 0.1$.

4. Inspections and Enforcement

The automatic use of an inspection system for enforcement is shown in Figure 3, which is equivalent to Model D of Kilgour (1994b). The analysis of this one-person game is straightforward. If (α, β) satisfies

$$\alpha F_S + \beta(P + T) < P \quad , \qquad (1)$$

then Subject always chooses Comply; if the reverse inequality holds, then Subject always chooses Violate. The transitional case, when equality holds in (1), is significant, and will be mentioned below. The set of (α, β) points where (1) holds is called the *C-Zone*

(Compliance Zone), and the set where the reverse inequality holds is the *V-Zone* (Violation Zone). The C-Zone and the V-Zone are shown in Figure 2.

The boundary between Compliance and Violation is a straight line that intersects the β-axis strictly between β = 0 and β = 1. When F_S is smaller than P, this line slopes gently downward. When P is smaller than F_S, it slopes steeply downward, with an α-intercept between 0 and 1.

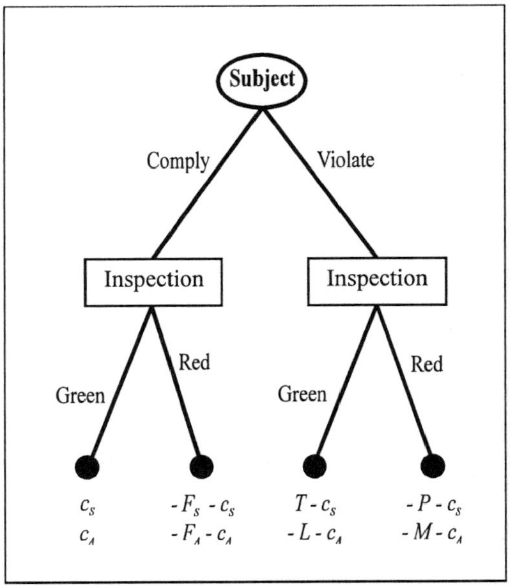

Figure 3: Automatic Enforcement

A particular binary sytem, with fixed (α, β), can be shifted into the V-Zone or the C-Zone by changing payoff parameters. This is illustrated by Figure 4 for the Example, using (α, β) = (0.3, 0.2). If P is small enough relative to T, (α, β) is in the V-Zone, and automatic enforcement results in 0% compliance. But a large enough value of P puts (α, β) into the C-Zone, where automatic enforcement yields 100% compliance.

In Automatic Enforcement with Imperfect Inspection, the boundary between 0% compliance and 100% compliance has equation

$$P = \left(\frac{\alpha}{1-\beta}\right) F_S + \left(\frac{\beta}{1-\beta}\right) T \quad .$$

If β > 0, there exist positive values of P and T for which automatic enforcement with (α, β) inspection gives 0% compliance, whereas no enforcement at all gives partial compliance - see Figure 4. In general, if P and T satisfy

$$T > \left(\frac{1-\alpha-\beta}{\beta}\right) F_S \quad \text{and} \quad F_S < P < \left(\frac{\alpha}{1-\beta}\right) F_S + \left(\frac{\beta}{1-\beta}\right) T \quad ,$$

then it is better to use no inspections at all than to use an (α, β) system automatically. On the other hand, the situation may be quite different if the (α, β)-value can be adjusted, as explained in Kilgour (1994b, Section 6).

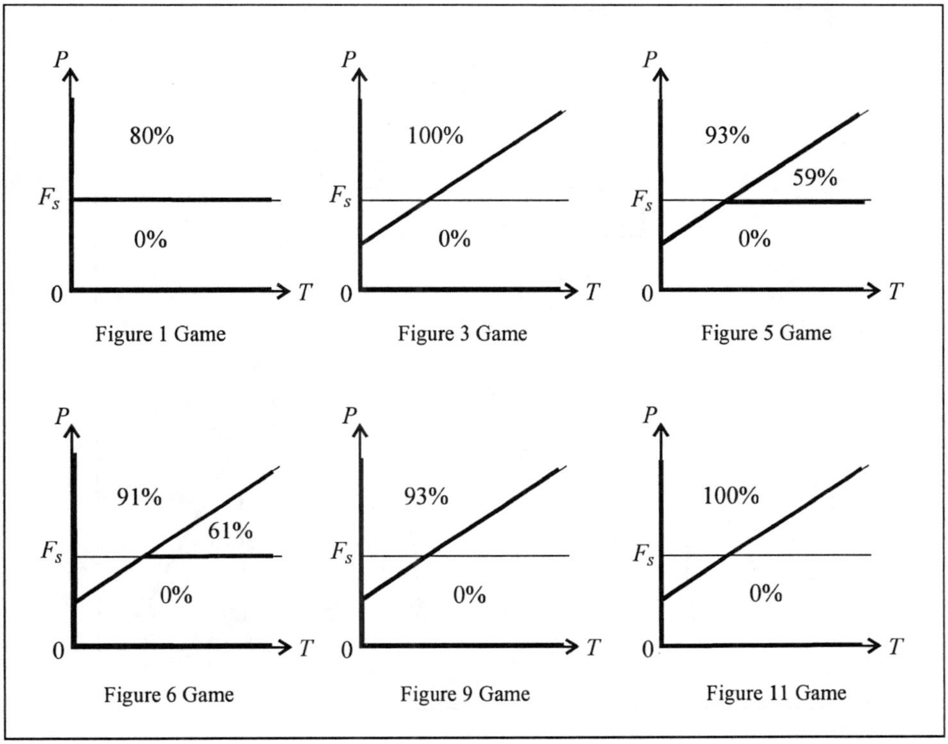

Figure 4: Compliance Measures for Enforcement Models - Example

One possible combination of the enforcement models of Figures 1 and 3 is shown in Figure 5. Note that Agency receives the signal from the inspection system, and must decide whether to Accept or Accuse based on knowledge of the game and the signal received. Of course, Agency may act probabilistically. Define the strategic variables

Subject: $x = Pr\{\text{Comply}\}$

Agency: $y_G = Pr\{\text{Accept} \mid \text{Green}\}$, $y_R = Pr\{\text{Accept} \mid \text{Red}\}$.

Provided $\alpha > 0$, the game of Figure 5 has only three possible equilibria (except in transitional cases), as follows:

(a) "Always Violate, Always Accuse"

$$x = 0; \; y_G = 0; \; y_R = 0; \; \text{iff} \; P < F_S \, .$$

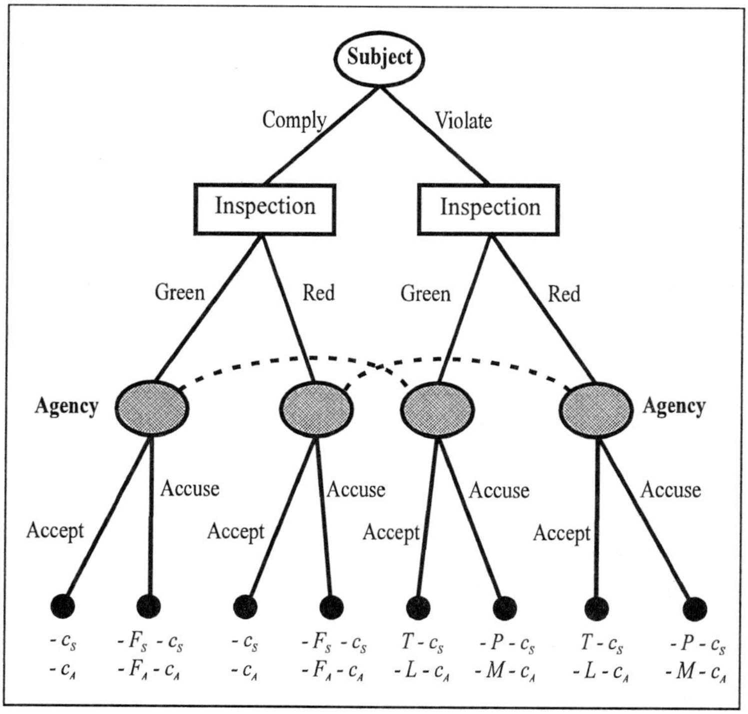

Figure 5: Agency Enforcement with Inspection

(b) V-Zone Equilibrium: "Sometimes Violate, Always Accuse if Red."

$$x \equiv x_b = \frac{\beta(L - M)}{(1 - \alpha)F_A + \beta(L - M)}; \; y_G = \frac{P - F_S}{\beta(P + T) - (1 - \alpha)F_S}; \; y_R = 0;$$

$$\text{iff} \;\; F_S < P < \left(\frac{\alpha}{1 - \beta}\right) F_S + \left(\frac{\beta}{1 - \beta}\right) T \; .$$

(c) C-Zone Equilibrium: "Sometimes Violate, Never Accuse if Green."

$$x \equiv x_0 = \frac{(1 - \beta)(L - M)}{\alpha F_A + (1 - \beta)(L - M)}; \; y_G = 1; \; y_R = \frac{(1 - \beta)P - \alpha F_S - \beta T}{(1 - \beta)P - \alpha F_S + (1 - \beta)T};$$

$$\text{iff} \;\; P > \left(\frac{\alpha}{1 - \beta}\right) F_S + \left(\frac{\beta}{1 - \beta}\right) T \; .$$

Equilibrium (b) occurs exactly when (α, β) lies in the V-Zone of Figure 2 and $P > F_S$, which implies that violation does not dominate compliance. Equilibrium (c) occurs if and only if (α, β) lies in the C-Zone of Figure 2, where $P < F_S$ is possible. For the example, the locations of the equilibria and the corresponding compliance frequencies are shown in Figure 4. The possibility that $\alpha = 0$ will generally be ignored here; this is justified by the ubiquity of trace amounts of substances in the environment, and the wide variations in their concentrations that occur naturally.

For any set of parameter values and any (α, β), at least one of the three equilibria occurs. Ignoring transitional cases, the equilibrium is unique, except in the triangle where

$$\left(\frac{\alpha}{1-\beta}\right) F_S + \left(\frac{\beta}{1-\beta}\right) T < P < F_S ,$$

where (a) and (c) overlap. It can be shown that Subject always prefers (c) to (a), while Agency prefers (c) to (a) if and only if $(1 - \beta) M > \alpha F_A$. This inequality is certainly true in the Example, and holds in general as long as F_A is not too large. This justifies the convention, adopted in Figure 4 and below, of showing only (c), the C-Zone equilibrium, in the region of overlap.

In the transitional case, where (α, β) falls on the boundary between the C-Zone and the V-Zone, equilibria (b) and (c) both occur, and have identical Agency strategies, $y_G = 1$, $y_R = 0$. As well, all convex combinations of (b) and (c) are equilibria, so Subject's equilibrium strategies are all compliance probabilities x satisfying $x_b \leq x \leq x_0$.

At this point, an important conclusion about the integration of an inspection system into an enforcement process becomes clear. The difference between the models proposed so far, as is evident from a comparison of Figures 3 and 5, is Agency's opportunity for "afterthoughts," i.e. for deciding whether to Accept or Accuse after receiving the signal generated by the inspection system. Inspection systems, even those with moderately high probabilities of error (subject to $\alpha + \beta < 1$) are more effective at inducing compliance when the action to be taken depends only on the technical result produced by the inspection system, and not on any subsequent choice by the enforcement agency. This point is discussed in detail in Kilgour (1994b).

Thus the enforcement procedure modeled in Figure 3 is better than that of Figure 5; as Figure 4 indicates, when circumstances are favorable, the former induces compliance always, and the latter only with high probability. Another possible problem with both approaches reflects agencys' inspection budgets. If compliance is to be induced, always or mostly, must an agency inspect for certain, thereby being obligated to pay the costs of an inspection that rarely yields evidence of a violation?

A procedure that avoids this problem by combining Figure 1 and Figure 3 is shown in Figure 6. Essentially, this procedure was proposed by Wittman (1989) in the context of arms control, and will be referred to here as Wittman Enforcement. In Wittman

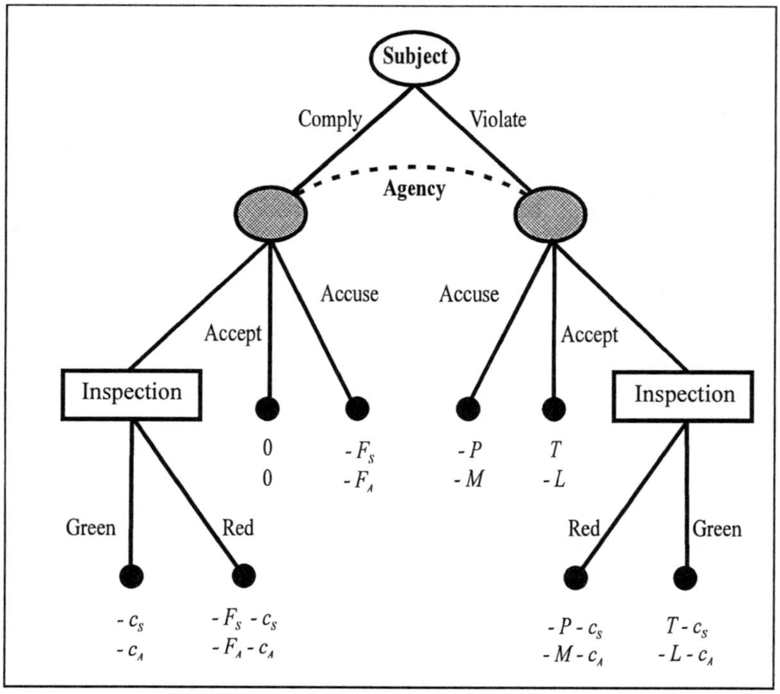

Figure 6: Wittman Enforcement

Enforcement, Agency selects one among three alternatives - either Accuse with no inspection, Accept with no inspection, or Inspect and act according to the signal received.

The solution of the game of Figure 6 is straightforward. Let the strategic variables y_P, y_S, and w denote the probabilities that Agency chooses to accept, accuse, and inspect, respectively. (Note that $y_P + y_S + w = 1$.) Provided $\alpha > 0$, all equilibria of the game of Figure 6 are as follows:

(a) "Always Violate, Always Accuse"

$$x = 0; \; y_P = 0; \; y_S = 1; \; w = 0; \;\; \text{iff} \;\; P < F_S \, .$$

(b) "Do Not Inspect"

$$x = \frac{L - M}{L - M + F_A}; \; y_P = \frac{P - F_S}{P + T - F_S}; \; y_S = \frac{T}{P + T - F_S}; \; w = 0;$$

$$\text{iff} \;\; \alpha + \beta > 1 - c_A \left[\frac{L - M + F_A}{(L - M) F_A} \right] \;\; \text{and} \;\; P > F_S \, .$$

(c) "Sometimes Accuse; Sometimes Inspect"

$$x = \frac{\beta(L - M) + c_A}{\beta(L - M) + (1 - \alpha)F_A}; \; y_P = 0; \; y_S = \frac{\beta T - (1 - \beta)P + \alpha F_S}{\beta(P + T) - (1 - \alpha)F_S}$$

$$w = \frac{P - F_S}{\beta(P + T) - (1 - \alpha)F_S};$$

$$\text{iff } F_S < P < \left(\frac{\alpha}{1 - \beta}\right)F_S + \left(\frac{\beta}{1 - \beta}\right)T \text{ and } \alpha + \beta \le 1 - c_A\left[\frac{L - M + F_A}{(L - M)F_A}\right].$$

(d) "Sometimes Accept; Sometimes Inspect"

$$x = \frac{(1 - \beta)(L - M) - c_A}{(1 - \beta)(L - M) + \alpha F_A}; \; y_P = \frac{(1 - \beta)P - \beta T - \alpha F_S}{(1 - \beta)(P + T) - \alpha F_S};$$

$$y_S = 0; \; w = \frac{T}{(1 - \beta)(P + T) - \alpha F_S};$$

$$\text{iff } P > \left(\frac{\alpha}{1 - \beta}\right)F_S + \left(\frac{\beta}{1 - \beta}\right)T \text{ and } \alpha + \beta \le 1 - c_A\left[\frac{L - M + F_A}{(L - M)F_A}\right].$$

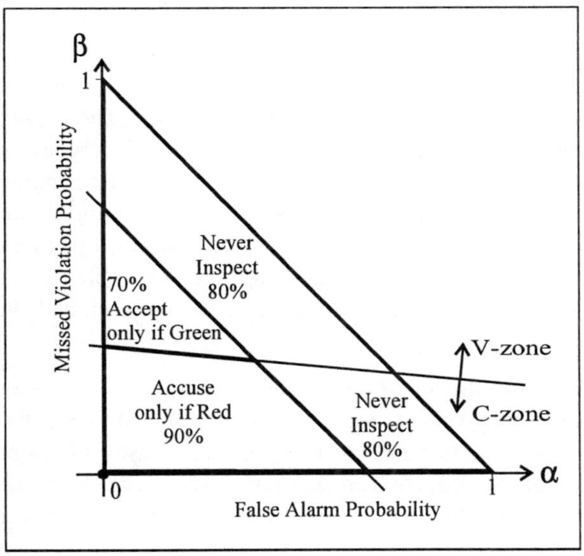

Figure 7: Equilibria in Wittman Enforcement

The locations of these equilibria in the (α, β) triangle of Figure 2 are shown in Figure 7, which is based on the assumption that $P > F_S$.

The four equilibria identified above are analogous to equilibria already uncovered; the elaboration of these analogies will make behavior under Wittman Enforcement easier to understand. Equilibrium (a) is the Violation-dominant equilibrium, at which Subject cannot be effectively deterred from violating. It appears in virtually all enforcement procedures and can be eradicated most easily by a sufficient increase in penalty.

Equilibrium (b) is precisely the equilibrium of the game of Figure 1, in which no inspection system is available. In fact, the first existence condition for equilibrium (b) is easy to interpret - if it holds, the inspection system is too expensive to be useful, and Agency is well-advised to save its resources and make do without any inspection signal. Rearrangement shows that this existence condition is equivalent to

$$c_A > \frac{(1 - \alpha - \beta)(L - M)F_A}{L - M + F_A} \quad .$$

The right side of this inequality provides a natural measure of the value of an inspection system, a measure that can obviously be compared to the system's cost to Agency. Note that the inspection system becomes more valuable as F_A and $(L - M)$ increase, and as α and β decrease - and that the value is equally sensitive to either increase and to either decrease. This expression constitutes a cost ceiling for inspection systems - more costly systems should simply not be used.

Equilibrium (c) is analogous to the V-Zone equilibrium in the game of Figure 5, and equilibrium (d) to the C-Zone equilibrium of that game. For instance, at the C-Zone equilibrium, there is always an inspection; if it produces a Green signal, the response is Accept, whereas if it produces a Red signal the response may be either Accept or Accuse with specified probabilities. Thus no accusations occur without a Red signal. Likewise in equilibrium (d) above, the action is Accept if either there is no inspection, or the inspection produces a Green signal; Accuse can be selected only if there is an inspection, and if that inspection produces a Red signal.

This analogy might suggest that the two most useful equilibria for the game of Figure 6 are essentially the same as the two important equilibria for the game of Figure 5. These pairs of equilibria are, however, not identical, as comparison of the resulting compliance levels indicates. In both cases, the highest achievable level of compliance is greater in the game of Figure 5. Although small in the case of the Example, the difference is evident in Figure 4. Compliance is greater at equilibrium (c) than at the V-Zone equilibrium, but these equilibria can hardly be recommended as methods for inducing compliance - as is clear from Figure 4, compliance would be higher with an enforcement system that included no inspections at all.

It should not be concluded, however, that the Figure 6 procedure is definitely worse than the Figure 5 procedure. In fact, Agency's payoffs are higher at equilibrium (d) than at the C-Zone equilibrium, reflecting that the reduction in Agency's costs as a result of fewer inspections more than compensates for the reduction in Agency's value because of the increase in violation rate - in other words, the benefit of Wittman Enforcement, fewer inspections, outweighs the cost, less compliance.

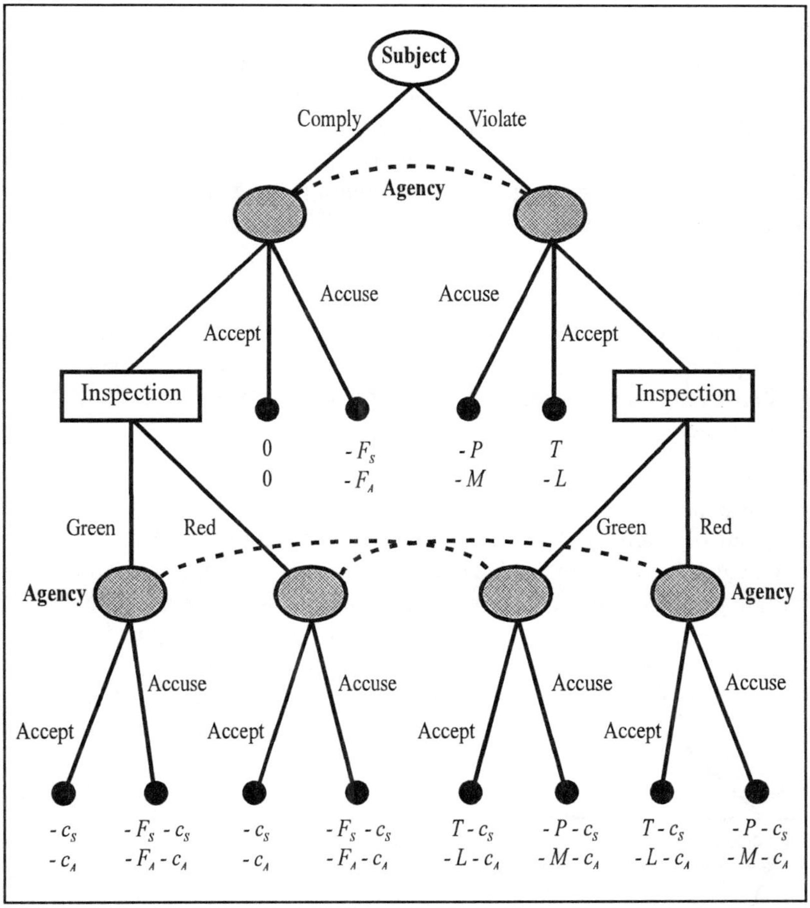

Figure 8: Wittman Enforcement with Subsequent Decision

This cost/benefit calculation highlights the dichotomy discussed in the Introduction. A good enforcement system keeps compliance high and costs low. But even when penalties are large enough to achieve the highest possible compliance rate, there is a conflict between enforcement effectiveness and enforcement costs. The concept of enforcement efficiency embodies this tradeoff. Which enforcement system achieves the greatest efficiency depends on the particular characteristics of enforcement problem under consideration.

It was noted earlier that giving Agency the opportunity to decide how to act based on the inspection signal is generally not a good idea. The addition of this subsequent decision, referred to above as "the possibility of afterthoughts," to Wittman Enforcement is illustrated in Figure 8. In this game, Agency first decides whether to Accept, Accuse, or Inspect; in the latter case, it then decides whether to follow up the inspection with the action "recommended" by the signal (Accept if Green, Accuse if Red).

Analysis of the game of Figure 8 shows that in fact nothing has been gained, although nothing has been lost either. The only Trembling-Hand Perfect equilibria are essentially duplicates of the equilibria of the game of Figure 6. In particular, if the inspection system is not very costly, and the penalty is high, then the equilibrium to be expected in the game of Figure 8 is equilibrium (d) of the game of Figure 6. In this instance, Agency's best strategy is to follow precisely the recommendation of the inspection system.

Allowing Agency to make decisions after it receives the inspection signal seems never to be worthwhile, so it will not be emphasized in the remainder of this paper. However it should be recognized that automatic conversion of the inspection signal into an action is an ideal that may not be attainable in many instances. Especially in matters as politically sensitive as environmental enforcement, it may be very difficult for authorities to find a credible ways to deny themselves the opportunity to make a decision.

Enforcement with a rule of evidence was introduced by Kilgour *et al.* (1993). A *rule of evidence* corresponds to the requirement that evidence of a violation be exhibited before any punishment can be applied. For instance, a prosecutor must usually submit "enough" evidence of a violation of the criminal law (at a preliminary hearing, for example, or before a grand jury) to justify a court's efforts to make a decision. In constrast, environmental laws usually give enforcement agencies the authority to impose control orders, which amount to penalties, with no intervening court process in which the evidence can be tested.

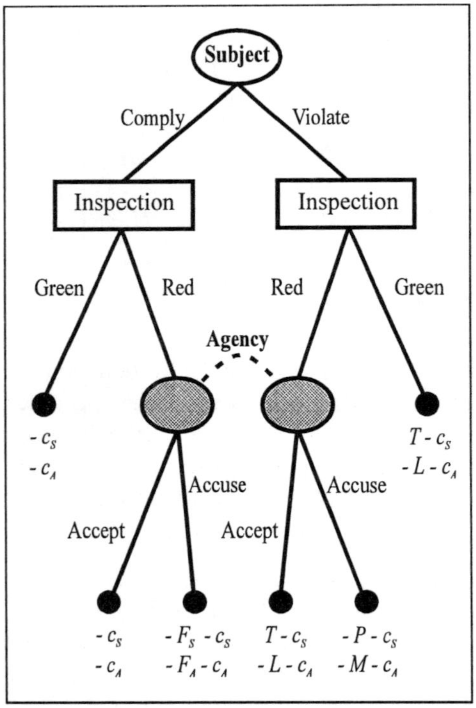

Figure 9: Rule of Evidence - Model 1

Rules of evidence are generally irrelevant when decisions to accuse or accept occur automatically, according to the signal produced by the inspection system. A rule of evidence can make a difference only in procedures in which Agency may either Accuse or Accept after an inspection signal, or with no inspection at all. The effect of the rule of evidence is that unless there is a Red signal (evidence of a violation), Subject cannot be accused.

Figures 9 and 10 show two different implementations of a rule of evidence for enforcement of a binary regulation using a binary inspection procedure. The game of Figure 9 is a simple modification of the game of Figure 5, denying Agency the opportunity to Accuse after a Green signal.

Note that the inspection is always conducted, and always paid for. The analysis of the game of Figure 9 makes clear its relation to the game of Figure 5. Equilibrium (a) ("Always Violate, Always Accuse") and equilibrium (c) (the C-Zone equilibrium) survive, while equilibrium (b) (the V-Zone equilibrium) disappears. Thus, as shown in Figure 4, only the best and worst cases remain. The loss of the rather unattractive intermediate possibility is the only consequence of moving from Figure 5 to Figure 9.

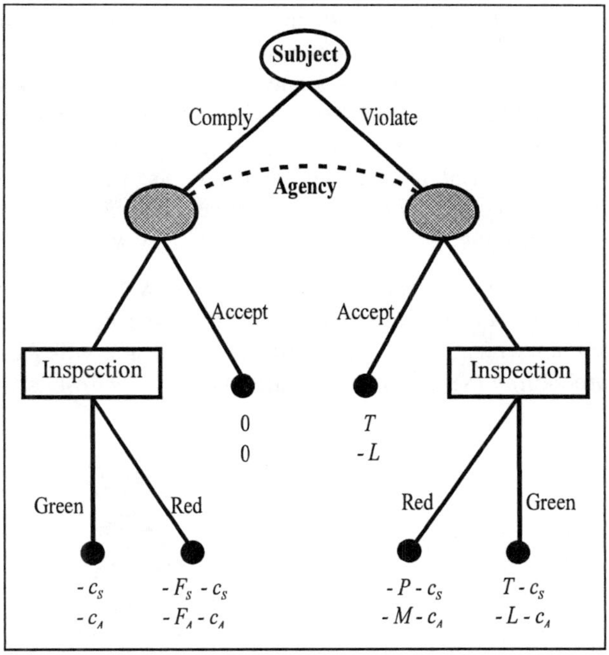

Figure 10: Rule of Evidence - Model 2

The situation is similar with the game of Figure 10, which bears a close relation to Wittman Enforcement (Figure 6) - except that Agency now cannot accuse without inspection. The equilibria of the game of Figure 10 are similarly related to the equilibria for the Wittman game - the V-Zone equilibrium disappears, and the properties of the other equilibria are unchanged.

Echoing the conclusions of Kilgour (1994b), both of these models suggest that a rule of evidence is innocuous if enforcement conditions are favorable. Only if penalties are relatively ineffective or inspections not too costly does the imposition of a rule of evidence alter the parties' behavior.

Another measure that has been suggested to improve the performance of enforcement systems is the use of policy precommitments by enforcement agencies. The idea is that prior announcements by an agency, if credible, may change a subject's behavior, and thus alter the way the system functions. Typically, the agency would announce its procedures for deciding its response to evidence it may receive in the future. This idea seems to have been proposed first in the context of arms control. Maschler's (1966) version of "inspector leadership" was developed further by Avenhaus (1990) and Kilgour and Brams (1992).

A common example of policy precommitment is the practice of advertising the numbers of speed traps, etc., to be operated by traffic police on a particular day. Many inspection agencies make public the details of where and how compliance data is to be collected. Simply put, Agency tells Subject: Faced with this information about how information will be obtained, analyzed, and acted upon, do you really want to violate?

Note that credibility of the precommitment is crucial. It may be advantageous for Agency to make announcements that, if believed, will change Subject's behavior, and then to renege - Agency gets no *immediate* benefit from following through on its announcements once Subject's behavior has been manipulated. We do not model questions of credibility here. We simply assume that Agency always carries out its commitments, perhaps because the enforcement interaction we model is part of a long-term or large-scale relationship, and Agency recognizes that failure to deliver on its commitments in any specific instance would jeopardize the future benefits that credibility would bring.

One possible implementation of Policy Precommitment is as Model G of Kilgour (1994b). In this case, it is known that Agency will inspect for certain; Agency's prior announcements concern the probability that it will Accuse after a Red signal, and the probability that it will Accept after Green. The results will not be reproduced here, but it

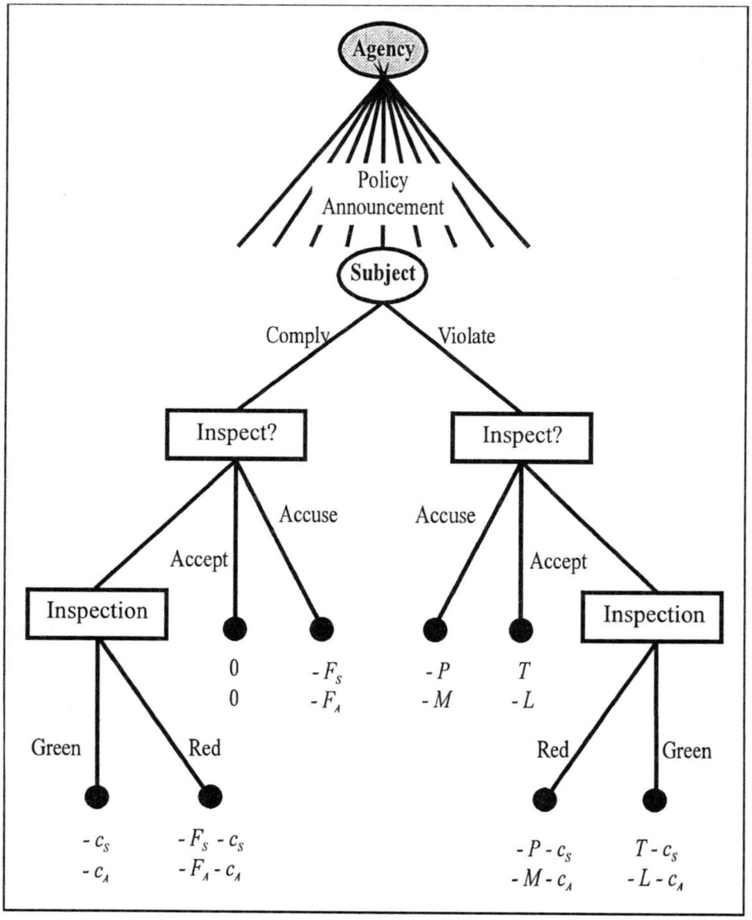

Figure 11: Enforcement with Policy Precommitment

should be noted that this system of precommitted responses is effective; 100% compliance is guarantee, subject to a few restrictions on the payoff parameters, and values of α and β in the C-Zone.

A different system of Policy Precommitment, more promising in light of observations made above, is shown in Figure 11. In this procedure, the precommitment is to whether there will be an inspection, and not to how the inspection signal is to be interpreted. This should be an improvement for two reasons: every inspection that must be paid for is in fact used, and action is automatic on receipt of signal, so there are no Agency "afterthoughts."

Analysis of the game of Figure 11 confirms the efficiency of this system. Under certain mild conditions on the parameters, plus C-Zone values of α and β and a cost-effective inspection process, the unique Nash equilibrium includes compliance for certain, along with a relatively high inspection frequency, as shown in Figure 4.

In summary, precommitments by Agency - credible prior announcements of how often it will inspect, and what it will do if it does not inspect - can be effective at achieving 100% compliance, provided the inspection system is good and other mild conditions are met. However structural requirements, including credibility of announcements and knowledge of payoff and inspection parameters, are crucial to the success of precommitments. The effectiveness and efficiency of the precommitment policy thus depends on its being implemented carefully and publicly in appropriate circumstances. Despite some restrictions, this is a very promising direction for future research. One recent development is the proposal by Fukuyama *et al.* (1994) of a *review policy*, essentially a detailed precommitment (analogous to tit-for-tat) about the collection and use of evidence.

5. Conclusions

Conclusions concerning expected compliance frequencies follow directly from the analysis of the models presented above, as do methods for measuring the temptation to violate. In addition, the models suggest how the maximum level of enforcement efficiency depends on the parameters of the model. Thus, the efficiencies of monitoring and inspection systems have been compared, the enforceability of regulations has been assessed, certain design issues for enforcement systems have been identified, and the usefulness of monitoring systems has been measured. Methods addressing the particular problems of environmental enforcement have been emphasized.

The inferences drawn here can be compared to other studies of how a reluctant decision-maker can be induced to behave in accordance with a rule. For instance, Russell (1990) provides a model of how a penalty and inspection policy can be linked to past behavior to maximize enforcement efficiency. As well, Fukuyama *et al.* (1994) suggest an enforcement policy that should induce compliance over the long term. The problems and methods of environmental enforcement are comparable with those in related fields, such as arms-control verification. It should be possible to develop these connections further. (Kilgour 1994a, b)

In summary, this study uses models of the enforcement of environmental regulations to obtain some guidance about how regulations can be designed, and how environmental enforcement agencies can function, so as to achieve acceptable levels of compliance as efficiently as possible. The conclusions represent a contribution toward the success of new rules governing human decisions, rules that will make it easier to attain and maintain a livable environment.

References

Andreoni, J. (1994). "Criminal deterrence in the reduced form: a new perspective on Ehrlich's seminal study," Public Choice Society, Austin TX.

Avenhaus, R. (1990). "Decision procedures for safeguards systems," *Information and Decision Technologies*, **16**, 229-247.

Canada's Green Plan (1990). Government of Canada, Department of Supply and Services, Ottawa ON.

Ehrlich, I. (1973). "Participation in illegitimate activities: A Theoretical and Empirical Investigation," *Journal of Political Economy,* **81**, 521-564.

Fang, L., Hipel, K.W., and Kilgour, D.M. (1994). "Enforcement of environmental laws and regulations: a literature review," *Effective Environmental Management for Sustainable Development,* (K.W. Hipel and L. Fang, eds.) Kluwer, Dordrecht, The Netherlands, 3-16.

Fudenberg, D. and Tirole, J. (1991). *Game Theory,* MIT Press, Cambridge MA.

Fukuyama, K., Kilgour, D.M., and Hipel, K.W. (1994). "Systematic policy development to ensure compliance to environmental regulations," *IEEE Transactions on Systems, Man, and Cybernetics,* **24**, 9, 1289-1305.

Galentine, C.A. (1994). "Taxpayers, examiners, and the government: a three-party principal-agent model," TIMS XXXII, Anchorage AK.

Hipel, K.W., Fang, L., Kilgour, D.M., and Haight, M., (1993). "Environmental conflict resolution using the graph model," *Proceedings,* IEEE International Conference on Systems Man and Cybernetics, Le Touquet, France, **1**, 153-158.

Keeney, R.L. (1994). "Using values in Operations Research," *Operations Research,* **42**, 5, 793-804.

Kilgour, D.M. (1994a). "Decision making in verification: the theory/practice interface," Report of the Laurier Centre for Military Strategic and Disarmament Studies, Wilfrid Laurier University, Waterloo ON.

Kilgour, D.M. (1994b). "The use of costless inspection in enforcement," *Theory and Decision,* **36**, 207-232.

Kilgour, D.M., Fang, L., and Hipel, K.W. (1993). "Game-theoretic analysis of enforcement of environmental laws and regulations," *Water Resources Bulletin,* **28**, 141-153.

Kilgour, D.M. and Brams S.J. (1992). "Putting the other side 'on notice' can induce compliance in arms control," *Journal of Conflict Resolution,* **36**, 3, 395-414.

Maschler, M. (1966). "A price leadership method for solving the inspector's non-constant-sum game," *Naval Research Logistics Quarterly,* **13**, 11-33.

Russell, C.S. (1990). "Game models for structuring monitoring and enforcement systems," *Natural Resource Modelling,* **4**, 143-173.

Russell, C.S., W. Harrington, and W.J. Vaughan (1986). *Enforcing Pollution Control Laws,* Resources for the Future, Washington DC.

Tsebelis, G. (1990). "Penalty has no impact on crime: a game-theoretic analysis," *Rationality and Society,* **2**, 225-286.

Wittman, D.A. (1989). "Arms control verification and other games involving imperfect detection," *American Political Science Review,* **83**, 3, 923-945.

VALUATION OF ENVIRONMENTAL GOODS

Frame effects in the estimation of multi criteria utility functions

FRED WENSTØP AND ARNE J. CARLSEN
Norwegian School of Management
P.O.Box 580, N-1300 Sandvika, Norway

Abstract

Health impacts of emissions from automobile traffic have been valuated with expert panels in a study reported elsewhere. To investigate possible biases created by framing effects, a separate study was conducted. 200 students were given a case describing city planning options in a small Norwegian city. They formed panels of maximum three persons each. They were expected to act as concerned citizens and try and build consistent multi criteria utility functions to determine how much they think the society should spend to alleviate various health problems connected with emissions from automobile traffic. To study possible framing effects, the students were randomly handed one of two possible cases. In the one case, all health impacts and costs were half the size of those in the other.

In the valuation process, the students used an interactive weight elicitation program based on pairwise trade-offs between criteria. To limit the impact of framing effects, the panels were further required to transform the criteria weights to monetary willingness to pay per impact unit of the criterion and use this as a basis for a reassessment of the weights.

The results show significant bias towards higher valuations of health impacts in the smaller city. The tendency is strong for some impact categories, but not all. Strong indication of the same kind of bias was also found in the main study where three expert panels from governmental agencies participated in a similar, but much more thorough process.

1. Introduction

The consensus is growing that environmental- and health impacts of public and private enterprises need to be assessed on equal terms with effects that can be measured directly in monetary terms through the price mechanism. Thus non-economic impacts should be taken into account in a consistent fashion during the project evaluation phase instead of as secondary effects. This can in principle be done by assigning *importance weights* (or prices) to goods that are not traded in the market.

In his book on environmental philosophy, Attfield (1994) calls for *Comprehensive Weighing*, a process whereby all relevant impact domains are valued so that the weights (or prices) are open to public discussions, thereby facilitating communication and debate about the proper weights to be used in a given decision situation.

Our work is a step in the direction of making such processes operational. Our aim is to investigate whether the methods of multi criteria decision analysis (MCDM) as outlined by Keeney and Raiffa (1976) can be advantageously employed for such purposes. Our concern is health- and environmental impacts from automotive traffic.

R. Nau et al. (eds.), Economic and Environmental Risk and Uncertainty, 53–64.

Specifically, we want to contribute towards methods whereby such impacts can be valuated through actual cases in a way which makes the results transferable to other projects.

The present report is part of a larger project to valuate health- environmental impacts from automotive traffic. In the main study, three expert panels within governmental agencies were interviewed to elicit their willingness to pay to lessen the burden of diverse impacts (Wenstøp et al. 1994). The study reported here was conducted with student panels and especially designed to investigate potential bias problems with the elicitation methods that were used in the main study.

1.1 POPULATION SURVEYS AND EXPERT PANELS

Most countries require today systematic *consequence analysis* of large scale public enterprises. In consequence analysis, the factual effects of the different decision alternatives are estimated. When, as is usually the case, the consequences are multi-dimensional, we have a multi criteria decision problem. Then the factual effects alone are not sufficient to rank alternatives. One must also take into account the importance of the different decision criteria. For goods that are traded in a free market, the market prices express the population's valuation of those goods. Environmental and health impacts are not traded, however. Such goods must therefore be valuated by other means. For lack of rational processes, many projects analyses stop short at this point and the actual decision is left to more haphazard political processes. What is needed, are tools which help the decision makers obtain values for the different impacts in a consistent fashion.

A well established approach is *contingent valuation* (CV). CVs are surveys where respondents state their *willingness to pay* for selected health- and environmental goods. The method has been described in reviews by Cumming et al. (1986) and Mitchell and Carson (1989). The approach has the merit of bringing forth "objective" values that are presumed to exist in the population. If we believe in democratic principles that stipulate that decisions should be made to reflect the values of the people, this is exactly the kind of information we need. Cowen (1993) refers to this as the *preference sovereignty norm.* He warns, however, that the norm is incoherent and that it "does not even have a positive marginal product in judging welfare or making social decisions". Other authors have discovered more pragmatic, but still troubling problems with the CV method. Brown and Slovic (1988) finds it susceptible to influence from cognitive and contextual biases. Kahneman and Knetsch (1992) have argued that CV responses denote moral sentiments rather than economic values. In a penetrating paper, Gregory et al. (1993) state that the major problem with CV is that preferences for environmental goods do not exist *a priori* in the minds of common citizens. Attempts to measure something that does not exist will lead to arbitrary biases and random effects. To increase validity, they advocate that preferences should be *constructed* through suitable processes. They examine techniques derived from multiattribute utility theory (MAUT) and decision analysis (DA) and recommend that such methods be used. We want to pursue this line with the help of the ideas originating with the seminal work of Keeney and Raiffa (1976).

To construct preference models requires interactive processes with the respondents. Large scale surveys of the population will therefore be rather costly. An obvious alternative is to work with a smaller number of expert panels instead. The panels would be appointed from relevant governmental agencies or other stakeholder groups. An early application is discussed by Wenstøp (1983) involving alternative oil spill combat strategies. One advantage of using expert panels is that their knowledge about the salient issues in the decision problem makes it easier to construct valid preference models (Bolger and Wright 1994). An argument against expert panels, on the other hand, is that the judgments of arbitrary expert panels lack legitimacy. That argument becomes weaker, however, if we look at experts, not as decision makers, but as *advisors* to democratically elected public servants. The important point is that the role of experts should be to bring forth qualified judgments in an understandable form that can be made open to scrutiny and discussion by anyone. This is precisely what Attfield has in mind with his *Comprehensive Weighing* paradigm. According to Stauth et al. (1993), weighting should be performed by expert panels within the responsible governmental agency. An alternative is to use independent expert panels to represent separate stakeholder groups. It makes it possible to discover important value differences that can be used to find alternatives that are good compromises.

1.2 DECISION ANALYSIS APPLIED TO ENVIRONMENTAL PROBLEMS

To perform value judgments is demanding. To be a good judge, you need to involve yourself in the matter you are dealing with and penetrate it. A real decision problem, as opposed to an artificially constructed case, has the advantage that reality is easier to grasp. When the outcome of the decision actually matters, the decision maker is more strongly forced to focus the mind on the important issues. To make the fullest use of the cognitive abilities of the experts, valuation of environmental goods should therefore be performed in connection with real decision problems.

Under the MCDM paradigm, a decision analysis usually starts with the construction of a goal hierarchy to define the relevant decision criteria. In environmentally related problems, the relevant criteria will by necessity include health- and environmental goods. In the next step, viable decision alternatives are identified followed by a consequence analysis. The results of that analysis can be summarized in the form of a *decision table* where the predicted scores of the criteria are listed for each alternative.

TABLE 1. A decision table gives the score x^i_k of criterion k, k=1,2..n when alternative i, i=1,2..m is chosen

	Alternative 1	Alternative 2
Criterion 1	x^1_1	x^2_1
Criterion 2	x^1_2	x^2_2

The value elicitation process usually starts with reference to the decision table. A valuation process usually ends up with weight estimates which reflect the importance of the decision criteria. In a decision analysis setting, the *importance* of a criterion depends both on its *intrinsic* value and *how much is at stake* in the decision problem at hand. The weight of a criterion will therefore be problem related, even though the intrinsic value may be the same from problem to problem. There exists a number of valuation schemes. Stewart (1992) gives a critical review of various methods. They fall broadly in two classes, those that are *decision problem oriented* and those that are *criteria oriented.*

Methods in the first class seeks expressly to exploit the experts' cognitive understanding of the decision problem by eliciting importance more or less directly. Popular examples are swing technique- and weight rating methods as discussed by von Winterfeldt and Edwards (1986). The methods work generally well in interaction with experts. Methodological problems are reported, however, that tend to bias the results. One such frequently observed bias is attributed to the framing of the problem, notably to the ranges of the criteria as they appear in the decision table. For consistency, to keep the intrinsic importance constant, the weight must increase if the range increases and *vice versa*. Von Nietzsch and Weber (1993) reports that the weight adjustments to range changes is insufficient when the weighting is performed in the decision problem context.

Methods in the second class focus on the decision criteria rather than on the alternatives. The corresponding weighting methods are generally called *trade-off* methods (Keeney and Raiffa 1976, Beinat 1995). With such methods, the ranges of the decision criteria are irrelevant and there no direct elicitation of the importance of the criteria. One therefore should expect less problems with framing biases.

In the present report, we study whether this holds true.

2. Criteria selection

The decision criteria represent the attributes or amenities to be valuated. They must be chosen carefully. Keeney (1992) advocates that they have the following properties:

- *Relevance*: A criterion should be a relevant and unambiguous indicator for the underlying objective it is meant to represent. For instance, if the underlying objective is to avoid loss in connection with fatalities, "the numbers of years lost" might be a more relevant criterion than "the number of lives lost" since it would also reflect the age distribution of the people that died.
- *Understandability*: Since the criteria are going to be weighted according to importance by the experts, it is paramount that they understand their ramifications. For instance, "a person exposed to above recommended limits of NO_2-concentration" is less understandable than "a cough-day" since it is harder to infer the consequential pain and suffering.

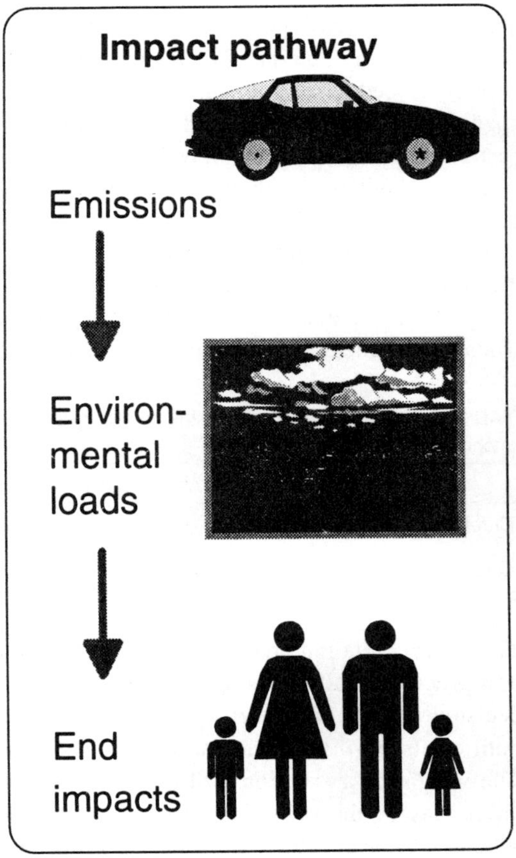

Figure 1. The impact pathway concept

- *Operationality*: When one analyzes alternative measures to alleviate adverse impacts, it must be possible to predict the effects of the alternatives measures with reasonable accuracy.

The more understandable a criterion is, the more *valid* the weight estimates will be. The more operational a criterion is, the more reliable the predicted effects of the alternatives in the decision problem will be. In practice therefore, understandability and operationality are conflicting objectives. Through discussions with our expert panels in the main study, we came to the conclusion that understandability is the more important concern. We therefore decided to use *end-impact* criteria rather than *environmental-load* criteria. It is obviously much easier to valuate a "person-day of chest irritation" than a "day of exposure to above recommended concentrations of NO_x". The disadvantage of using end-impact criteria, on the other hand, is that they are less *operational*. It is harder to use them for planning purposes, since it will make it necessary to make predictions further down the impact pathway. See figure 1.

Our conclusion is based on the belief that rising environment and health concerns will make valuing the external effects of alternative decisions in public planning more and more important, and that this can only be done in a reliable way if we focus on end-impacts. This thinking is in line with the procedure adopted in the CEC/US Joint Study on Fuel Cycle Costs (1993). Here a major emphasis has been placed on defining a set of well defined fuel cycle end impacts so that these can be valued and the values used in a consistent fashion in different countries. Our definition of end impacts in the present study, are as close as possible to those recommended in the CEC/US study. We have only made changes when the availability of data made it necessary. Table 2 shows the selected attributes to be valuated. Some units are in 1000s for practical reasons. "Haze" pertains to a specific town.

TABLE 2: Definition of the environmental impact criteria

Coughing	Decrease in # cough days in children	1000 child-day/year
Asthma	Decrease in # days with asthma in children and elderly	1000 child-day/year
Life years	Decrease in loss of years of life because premature deaths	year
Noise	Decrease in exposure to background traffic noise > 55DBA	person
Top Noise	Decrease in exposure to intermittent traffic noise > 70DBA	person
Smell	Decrease in # people complaining about bad smells	1000 personday/year
Haze	Decrease in # days with visible haze	day/year
Fear	Decrease in # children kept indoors for fear of traffic	child
Accidents	Decrease in # major injuries in traffic accident	injuries/year
Costs	Social economic costs incurred (investments and running)	NOK mill/year

3. Weight elicitation methods

Several different weight elicitation schemes have been computer implemented, and there are several reports comparing different implementations (Larichev et al. 1993). In the main study, we evaluated closely three alternative weighting schemes.

Ranking of alternatives is a holistic method where the experts rank an extensive list of alternatives in a decision table. Ordinal regression analysis will the produce weight estimates. It is implemented in MINORA (Hatzinakos et al. 1991).

Pairwise ordinal comparisons of criteria is a decision problem oriented method. It is implemented in DEFINITE (Jansson 1992) and leads to a ranking of the importance of the criteria based on verbal comparisons of pairs of criteria.

Pairwise numerical comparisons of criteria is a tradeoff method that requires indifference judgments from the experts. See table 3. It is described in the seminal book by Keeney and Raiffa (1976) and Beinat (1995) and implemented in *Pro&Con* (Gottschalk and Wenstøp 1988).

We ruled out ranking of alternatives since we want to use between 10 and 20 environmental impact variables as criteria. A ranking of alternatives with that many criteria was deemed to be too taxing for the decision maker. In the main study, we decided to use both pairwise ordinal and pairwise numerical comparisons. Since the method with pairwise ordinal comparisons is known to be liable to framing effects, we wanted to control the results with numerical comparisons although that method requires rather high precision from the panels. In this study, we investigate whether the method with numerical comparisons is biased as well. A brief discussion of the method follows after we have defined the concept of criteria weights.

3.1 CRITERIA WEIGHTS

Assume that we are given a decision problem. There are *n* decision criteria, $x_1,..x_n$ to be used to evaluate *m* alternatives. x^i_k is the score of criterion *k* when alternative *i* is chosen. x^i_k is restricted by its *range* where x^o_k is the least preferred value and x^*_k the most preferred. The range is derived from the decision problem.

Assume further that for each criterion a utility function has been specified so that $u_k(x^o_k)=0$ and $u_k(x^*_k)=1$. Given the decision problem, the *importance* of criterion *k* is represented by the weight w_k, $0<w_k<1$, $\Sigma w_k = 1$. The utility of a decision alternative with scores $x_1,..x_n$ is computed as:

$$U(x_1,..x_n) = w_1u_1(x_1)+w_2u_2(x_2)+..+w_nu_n(x_n) \tag{1}$$

From (1), it is obvious how the weights should be interpreted: w_k is the *marginal increase in the total utility* when the score of criterion *k* changes from x^o_k to x^*_k. Thus, the weight of a decision criterion has only meaning in relation to its range. Since the ranges are related to the decision problem, the weight of a criterion will reflect its importance *within the context of that decision problem.*

3.2 PAIRWISE NUMERICAL COMPARISONS OF CRITERIA

The weight elicitation process in Pro&ConW is a stepwise procedure:

- The expert panel is presented with a decision table $[x^i_k]$ describing a decision problem. The table lists the decision criteria and provides estimates of their scores for each decision alternative.
- The panel is asked to rank the criteria according to their importance in the decision problem.
- The panel is asked to specify a utility function for each criterion. The default is linear.

- The panel is asked to select one criterion as a reference criterion against which all the other criteria are going to be compared. This criterion should be easy to comprehend and it should not be monetary.
- The panel is presented with a two-by-two matrix and asked to adjust the numbers so that the "before" and "after" situations are equally preferable. All four numbers can be adjusted. See table 3.
- The relative weights are computed and the weights scaled to make their sum equal to one.
- The willingness to pay for one unit of each criterion is computed from the weights. This serves as a feedback in monetary terms, and the panels are invited to reassess the weights if deemed appropriate.

TABLE 3. A typical question posed in the pairwise comparison of criteria process

	Situation A	Situation B
x_i Coughdays p.a.	$x_i' = 750{,}000$	$x_i'' = 700{,}000$
x_r Public expenditures p.a.	x_r' = NOK 1.5 mill.	x_r'' = ?

Table 3 shows the kind of questions put before the panel. Two situations, A and B, are being compared. There are different scores on one criterion in the two situations, and a missing score on the other criterion in situation B. The panel is supposed to enter a number for the missing score that makes the two situations equally preferable, all other things being equal. Clearly, the public expenditures in B must be higher than NOK 1.5 mill. to compensate for the improvement in cough days. The exact amount will depend on the preferences of the panel. The initial numbers presented are chosen by the program so as to reflect the kind of trade-off problems inherent in the decision problem at hand. This way, the panel may apply any insight they may have achieved by contemplating the real decision problem. But as we shall see, this amounts to framing the valuation problem in a way which affects the tradeoff judgments.

4. The experiment

A class of students was given a case describing city planning options in a small city. The case was based on concurrent planning efforts in the Norwegian city of Drammen. It is described below. The students were supposed to form panels of maximum three persons who should act as concerned citizens and try and build consistent multi criteria preference models to determine how much they think the society should spend to alleviate various problems connected with emissions from automobile traffic. *To study possible weight range effects, the students were randomly handed one of two possible cases.* In one case (L) all health impacts and costs were *half the size* of those in the other (H). The panels were supposed to work independently and there was a strong incentive against collaboration. The panels used Pro&ConW as the preference construction tool.

4.1 ASSIGNMENT

> *"The authorities in a Norwegian city with approximately 100,000 inhabitants are evaluating seven different development strategies, all them targeted at alleviating environmental impacts from road traffic. Several regions of the city*

are today heavily polluted by noise and other emissions from petrol powered engines.

The seven alternatives are:

A - 50% of the petrol fueled cars will be substituted with el-powered ones

B - Cars are banished from the urban center

C - A toll ring around the city will decrease the number of vehicles in the city

D - A new ring road around the city will decrease the number of vehicles passing through

E - A major improvement of the collective transport system

F - A tunnel under the city to take care of the through traffic

G - Redirecting of trucks to prevent them from passing through the city

The estimated consequences of the different measures are shown in table 4. The definition of the impact criteria is shown in table 2. Your task is to assume the role of a concerned citizen and evaluate the different scenarios using sound and proper judgment. You are not living in the city yourselves, but you feel a natural concern for the well-being of your fellow countrymen.

To do:

Construct a multi criterion additive utility function which represents the preferences of the group as well as possible. Compute the implicit monetary value of one unit of each one of the criteria together with an interval to express your uncertainty about your judgment. Make a self evaluation of the robustness of the results and of how well you think they reflect your real preferences. Explain how you dealt with possible problems due to nonlinear utility functions."

TABLE 4: Decision table listing the consquences of seven alternative city development strategies. This table defines the "high" scenario (H). In the "low" scenario (L), all numbers are half the size.

Criterion	A	B	C	D	E	F	G
Coughing	500	750	350	750	250	550	400
Asthma	70	150	35	100	20	50	25
Life years	25	50	15	35	5	20	10
Noise	25000	-5000	-2500	0	10000	20000	17500
Top Noise	12500	10000	5000	25000	5000	10000	7500
Smell	3000	-500	500	2250	500	1500	1750
Haze	50	-25	-12,5	-10	20	30	-10
Fear	0	2500	1250	1000	500	1500	1000
Accidents	-5	25	5	15	10	20	12,5
Costs	3,5	15,0	1,5	10,0	2, 0	12,0	2,5

4.2 RESULTS

Of 94 panel reports, some panels had to be discarded from the data because of lack of clear results or obvious miscalculations or misunderstandings. This left us with 69 observations of valuations of the eight decision criteria. The valuations are stated in Norwegian crowns for one unit of the criteria. Some of the panels used nonlinear utility functions. In those cases, the valuations are average values.

The raw data are highly skewed for all criteria, with long right tails. This is common for this kind of data, where negative numbers are ruled out because of the context. Another contributing factor to the skewedness, is that the valuation of a criterion usually will be calculated by dividing the weight on the criterion by the weight on costs. Since both of these weights can be considered as random variables which may be normally distributed, we should expect a distribution of the outcome which resembles the lognormal distribution. Normal probability plots of the logarithms of the valuations show this to be a very good approximation.

TABLE 5: Geometric mean of all valuation responses with 95% confidence limits. Valuation in NOK (7 NOK = 1 USD), *n*=69

Valuation, geometric mean	**Cough** (person-day)	**Asthma** (person-day)	**Life** (person-year)	**Noise** (person-year)	**Topnoise** (person-year)	**Smell** (person-year)	**Fear** (child-year)	**Accidents** (personal injury)
average	22	110	290,000	310	350	2.05	3300	590,000
lower limit	17	86	225,000	245	280	1.5	2600	455,000
upper limit	28	145	375,000	390	440	2.8	4300	765,000

TABLE 6: Geometric mean of all valuation responses with 95% confidence limits. Valuation in NOK (7 NOK = 1 USD). The valuations are split on the High and Low scenario. The *t* and *p*-values are based on a *t*-test for the difference between two means.

		n	Average	Lower limit	Upper limit	t	p
Cough-H	person day	37	16	11	23	-2.72	0.01
Cough-L		31	32	23	44		
Asthma-H	person day	38	83	57	119	-2.75	0.01
Asthma-L		31	161	117	222		
Years-H	person year	38	264000	184000	379000	-0.81	0.42
Years-L		30	326000	225000	472000		
Noise-H	person year	37	287	209	394	-0.75	0.46
Noise-L		31	341	245	474		
T-Noise-H	person year	37	325	244	433	-0.67	0.5
T-Noise-L		31	381	263	553		
Smell-H	person year	38	2.2	1.4	3.4	0.5	0.62
Smell-L		31	1.9	1.2	2.9		
Fear-H	child year	38	2720	1980	3730	-1.75	0.09
Fear-L		29	4290	2840	6500		
Accid.-H	personal injury	37	438000	322000	597000	-2.58	0.01
Accid.-L		30	847000	564000	1272000		

Since the logarithms of the results follow approximately normal distributions, we report the results in terms of geometric means. The calculated means are shown in table 5 with 95% confidence intervals. If the data are further split according to the two cases, we see a rather strong tendency to a higher valuation in the low case for most criteria. This can be seen from table 6. The data in table 6 are plotted in figure 2.

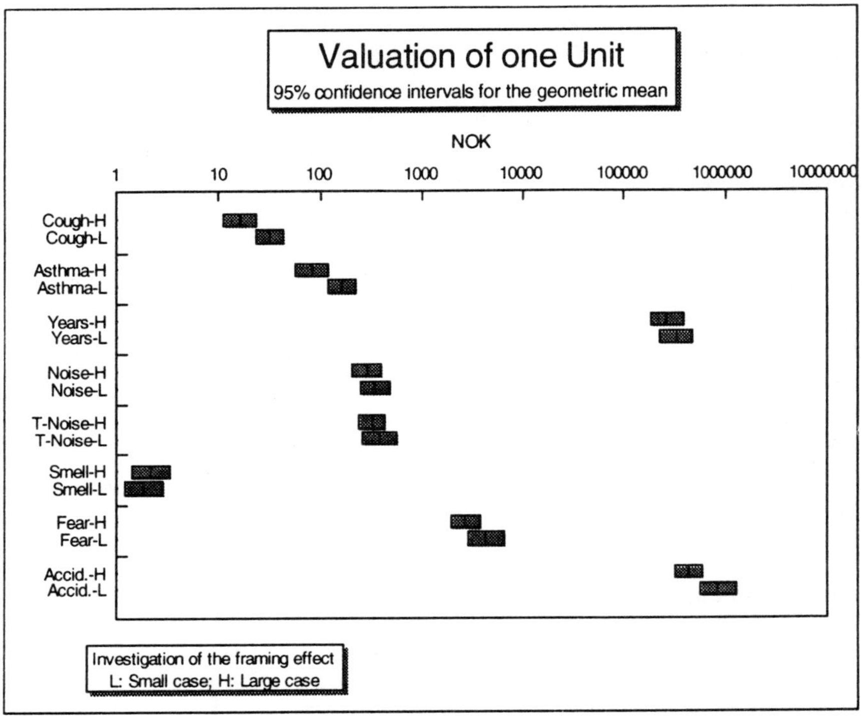

Figure 2. Evidence of the framing effect

To check whether the valuations are significantly influenced by the scenarios, we have performed *t*-tests on the means of the logarithms. The tests indicate significant framing effects from the scenarios for Coughing days, Asthma days, Accidents and possibly Anxiety for children. There is no indication of a framing effect for Years of life lost, Noise-years, Top-noise years or Smell years. We have as yet no explanation why the criteria come out differently in this test. Except for Smell, all criteria tend to be valued higher in the small case with limited consequences (L), than in the bigger case with twice the impacts. This is consistent with what was indicated in the main study with three expert panels from governmental agencies.

5. Discussion

To point at a psychological explanation for the framing effect, it is necessary to know the essence of the valuation process. Its core is the selecting of a value for one criterion to make two pairs of criteria values equally attractive. For instance, to judge the tradeoff between coughing days and costs, the panel is typically presented with a choice between A and B in the "High" scenario:

A: 750,000 coughdays and NOK 1.5 mill. in public expenditures

B: 700,000 coughdays and NOK *x* mill. in public expenditures

The panel must then judge how much will *x* have to be to make A and B equally attractive. Let us say the answer is *x*= NOK 2.5 mill. This makes the value of one coughday equal to NOK 20. The corresponding question for the "Low" scenario would be:

A: 375,000 coughdays and 750,000 NOK in public expenditures

B: 350,000 coughdays and *x* mill. NOK in public expenditures

We have shown that in independent valuations, the panels with the low scenarios tend to select *x* to make the value of one coughday significantly higher than the panels with the high scenarios. This phenomenon occurs even though the panels actually do the computation of the value of one unit of the criterion and use this information to review their responses.

A possible explanation is that the panels are influenced by imaginative budget restrictions which makes it disproportionately easier to spend smaller than larger amounts of money.

6. References

Attfield, R. (1994) *Environmental Philosophy, Principles and Prospects*, Avebury, Aldershot, UK

Beinat, E. (1995) *Multiattribute value function for environmental management*, Tinbergen Institute Research series 103, Vrije Universiteit, Amsterdam

Bolger, F. and Wright, G. (1994) Assessing the quality of expert judgments, *Decision Support Systems* **11** 1-24

Brown, T.C. and Slovic, P. (1988) Effects of Context on Economic Measures of Value. In Peterson, G. et al. (eds.) *Integrating Economic and Psychological Knowledge in Valuations of Public Amenity Resources*, State College, PA: Venture, 23-30

CEC/US Joint Study on Fuel Cycle Costs, Assessment of the External Costs of the Coal Fuel Cycle, 1993, Draft paper prepared for DG XII of the Commission of the European Community

Cowen, T. (1993) The scope and limits of preference sovereignity, *Economics and Philosophy* **9**, 253-269

Cummings, R. G., Brookshire, D. S. and Schulze, D. (1986) *Valuing Environmental Goods: Assessment of the Contingent Valuation Method*, Rowman ans Allanheld, Totowa NJ.

Gottschalk, P. and Wenstøp, F. (1988) *Quantitative Decision Analysis*, Part II, *Preference Analysis* (in Norwegian), Universitetsforlaget, Oslo

Hatzinakos, I., Yannakopolous, D., Faltsetas, C. and Ziourkas, C. (1991) Application of the MINORA decision support system to the evaluation of landslide favourability in Greece, *European Journal of Operation Research* **50**, 60-75

Jansson, R. (1992) *Multiobjective Decision Support for Environmental Management*, Kluwer Academic Publishers, Dordrecht, Holland

Kahneman, D. and Knetsch, J. (1992) Valuing Public Goods: the Purchase of Moral Satisfaction, *Journal of Environmental Economics and Management* **22**, 57-70

Karni, R., Feigin, P. and Breiner, A. (1991) Multicriterion issues in energy policy making, *European Journal of Operations Research* **56**, 30-40

Keeney, R. and Raiffa, H. (1976) *Decisions with Multiple Objectives,* John Wiley & Sons

Keeney, R. (1992) *Value-focused thinking,* Harvard University Press, London

Larichev, O.,Moshkovich H., Mechitov, A. and Olson, D. L. (1991) Experiments comparing qualitative approaches to rank ordering of multiattribute alternatives". *J of Multi-Criteria Decision Analysis* **2**, 5-26

Lawrence, D.P. (1993) Quantitative versus qualitative evaluation: A false dichotomy? *Environmental Impact Assessment Review* **13**, 3-11

Mitchell, R. C. and Carson, R.T. (1989) *Using Surveys to value Public Goods: The Contingent Valuation Method*, Resources for the Future, Washington, DC

Stauth, R., Sowman M. and Grindley S. (1993) The panel evaluation method: An approach to evaluating controversial resource allocation proposals. *Environmental Impact Assessment Review* **13**,13-35

Stewart, T.J. (1992) A critical survey on the status of multiple criteria decision making theory and practice. *OMEGA Int. J. of MGMT Sci.* **20**, 59-586

von Nitzsch, R. and Weber, M. (1993) The Effect of Attribute Ranges on Weights in Multiattribute Utility Measurements, *Management Science*, vol. 39, No. 8, 937-941

von Winterfeldt, D. and Edwards, W. (1986) *Decision analysis and behavior Research*, Cambridge University Press, Cambridge

Vuk, D., Kozelj, B. and Mladineo (1991) Application of multicriterion analysis on the selection of the location for disposal of communal waste, *European Journal of Operations Research*, **55**, 211-217

Wenstøp F. et al. (1994) *Valuation of environmental Goods with Expert Panels*. Norwegian School of Management Research Report 1994/1, Oslo

Wenstøp, F. (1983) Evaluation of Oil Spill Combat Plans by Means of Multi-Criteria Decision Analysis, in Stigum, B. and Wenstøp, F., (eds.), *Foundation of Utility and Risk Theory with Applications*, D. Reidel Publ. Co., Dordrecht, Holland

STOCHASTIC TREES AND MEDICAL DECISION MAKING

GORDON B. HAZEN
IE/MS Dept., Northwestern University
Evanston, IL 60208-3119
USA

JAYAVEL SOUNDERPANDIAN
University of Wisconsin-Parkside
Kenosha, WI 53141-2000
USA

1. Introduction

Medical decision problems that confront a patient often involve chances of serious health consequences that may extend many years into the future. Conventional decision tree diagrams are not well suited to represent such continuously distributed temporal uncertainties. As a modeling aid for such problems, Hazen (1992) introduced the *stochastic tree*, a diagram which combines features of decision trees and stochastic-process transition diagrams. Stochastic tree diagrams can not only depict continuously distributed temporal uncertainties, but, like decision trees, can also be rolled back to determine optimal decisions. This rollback procedure is an instance of the method of successive approximations from stochastic dynamic programming. It can be applied not only for risk-neutral objective functions, such as mean quality adjusted life years (QALY), but also for the calculation of expected utility (Hazen and Pellissier 1994) leading to quality and risk adjusted life years or QRALY.

The purpose of this paper is to introduce stochastic trees and discuss techniques (such as factoring, Coxian background mortality, risk-sensitive preferences, continuous-risk utility assessment and cost-benefit analysis) for their application to medical decision making.

R. Nau et al. (eds.), Economic and Environmental Risk and Uncertainty, 65–74.

2. Stochastic Trees

A stochastic tree is a transition diagram for a continuous-time Markov chain, unfolded into a tree structure. Stochastic trees have been used as modeling tools for the analysis of medical treatment decisions (Hazen 1992, 1993; Chang, Pellissier & Hazen 1993).

As an example, consider the stochastic tree in Figure 1, a simplified model of nonsurgical treatment of transient ischemic attacks, motivated by Matchar and Pauker (1986). The usual conventions for transition diagrams apply: The nodes represent states (Well, Stroke, Post Stroke) and the arrows represent transitions between states. The Dead state does not explicitly appear, being modeled by termination of the process. There are two types of arrows, corresponding to the two types of possible transitions. Wavy arrows are labeled with rates, and they signify transitions which take time to accomplish. For instance, from the initial state Well, there is an exponentially distributed duration with rate μ_0 until death, and an exponential (λ_s) duration until transition to Stroke. The next state occupied is determined by the shorter transition time, just as in any continuous-time Markov chain. We call nodes from which wavy arrows emanate *stochastic* nodes. Straight arrows are labeled with probabilities, and signify immediate transitions. Nodes from which straight arrows emanate play the same role as chance nodes do in a decision tree. Hazen and Pellissier (1994) discuss equivalent representations for stochastic trees and methods to simplify their structure. Hazen (1992) presents a complete stochastic tree representation of the Matchar and Pauker model.

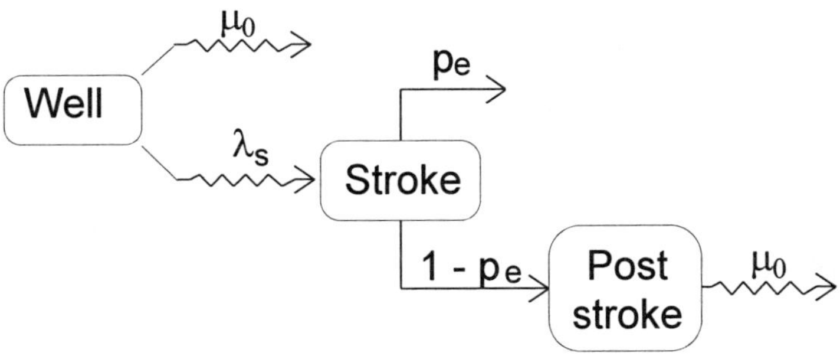

Figure 1. Stochastic tree model of transient ischemic attacks.

One popular measure of treatment efficacy used in the medical decision making literature is *mean quality-adjusted life years* or *mean QALY.* This measure is calculated by weighing each interval of time spent in a particular health state x by a *quality rate* v(x) proportional to the desirability of that state. So if T_X is the total duration spent in state x, then one seeks the mean value of $Q = \Sigma_X v(x)T_X$. Typically the Well state is assigned a quality factor of 1, the dead state 0, and other states intermediate values. QALY has been used in the analyses of hypertension (Weinstein and Stason 1976), brachycardia-tachycardia syndrome (Beck and Pauker 1981), osteoporosis (Hillner, Hollenberg and Pauker 1986), pyriform sinus carcinoma (Plante, Piccirillo and Sofferman 1987), prostatic cancer (Roach *et al.* 1988) and multiple sclerosis (Mooney, Mushlin and Phelps 1990).

A further advantage of the stochastic tree model is that it allows the recursive computation of mean QALY by "rolling back" the stochastic tree (Hazen 1992), much as one would roll back a decision tree. Consider a subtree of a stochastic tree in which an initial state x is occupied until one of several competing transitions to states y with rates λ_y occurs:

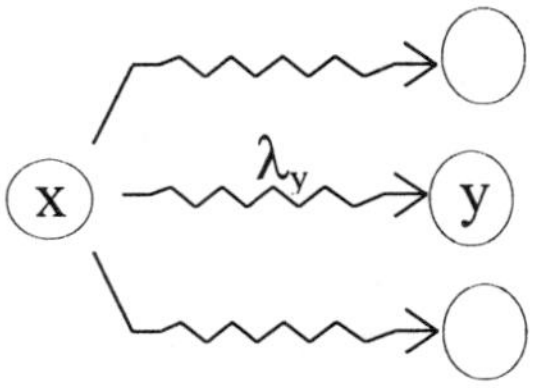

Suppose the mean QALY beginning at y is L(y). It is easily shown that QALY beginning at x is

$$L(x) = \frac{v(x) + \sum_y \lambda_y L(y)}{\sum_y \lambda_y}. \qquad (1)$$

Mean QALY can be recursively evaluated in any stochastic tree by taking the expectations at chance nodes and using (1) at stochastic nodes. If the tree contains directed cycles (repeated visits to a state) this formula will lead to iterative calculations. We have found it convenient to implement such iterative calculations on spreadsheets.

3. Factored Stochastic Trees

Quite often, a complex medical situation can be thought of as several health processes unfolding in parallel. For example, the model of Figure 1 contains a "background mortality" process with mortality rate μ_0 and, in parallel, a stroke process with stroke

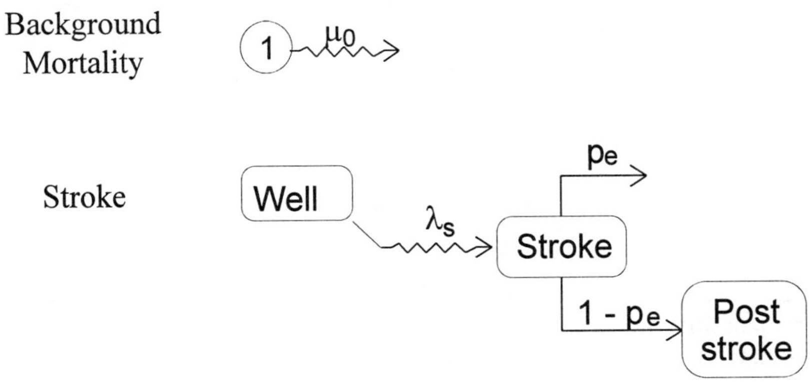

Figure 2: Factored representation of the stochastic tree in Figure 1

rate λ_s and stroke mortality p_e. We can depict these two processes as separate stochastic trees, as is shown in Figure 2. We then say we have *factored* the original tree into a background mortality factor and a stroke factor. Given a factored stochastic tree, the full tree can be constructed by forming the "Cartesian product" of the states in the background mortality factor with the states in the stroke factor, with each product state inheriting all transitions out of the factor states that compose it

This factoring approach facilitates modeling by allowing the modeler to focus separately on different problem aspects in order to refine or modify them as required. For example, the background mortality factor above assumes the lifetime to have an exponential distribution with mean μ_0, and this assumption can be refined. A background mortality factor which more accurately depicts human lifetime distributions is the Coxian model shown below:

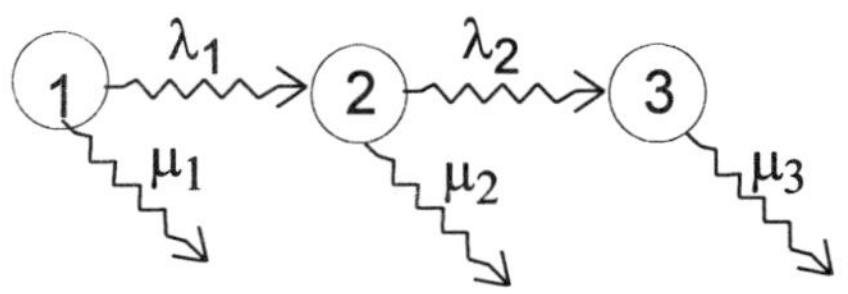

Here the individual may be thought of as proceeding through three "life stages" i = 1, 2, 3 where at each stage i the mortality rate is μ_i. The parameters λ_i, μ_i and the number of stages can be chosen so as to accurately approximate human survival distributions. The combination of this Coxian mortality factor with the original stroke factor is therefore an improved model of stroke. The Cartesian product of these two factors is the product tree:

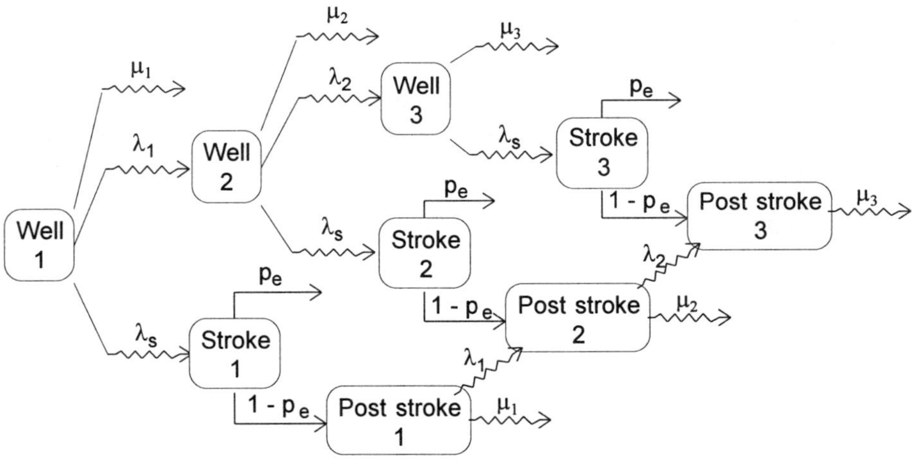

It is clearly easier to work with the two factors separately. To summarize, the factor approach simplifies model formulation by decreasing the chance of modeling errors, and improves model presentation by allowing a simpler display. Multifactor trees arise naturally. For example, Hazen (1993) presents a model with five factors representing background mortality, systemic embolism, pulmonary embolism, systemic hemorrhage and anticoagulant status.

4. Utility Functions Over Stochastic Trees

To consider preferences under non-neutral risk attitudes, there exist utility functions over stochastic trees that can be calculated by rollback (Hazen and Pellissier 1994). One such class of utility functions, with which we shall work for the rest of this paper, can be introduced as follows: Define a *history* $h = x_1^{t1}x_2^{t2}...x_n^{tn}$ as a finite sequence of states x_1, x_2, ..., x_n and durations t_1, t_2, ..., t_n, where it is understood that state x_1 is first occupied for duration t_1, followed by state x_2 for duration t_2, and so on. Recursively define a utility function u(h) over histories $h = x^t k$ with initial sojourn t in state x by

$$u(h) = \int_0^t v(x)e^{-a(x)s}ds + e^{-a(x)t}u(k) \tag{2}$$

where u(k) = 0 whenever k is null (i.e., when h is simply x^t). The quantity v(x) is a *quality rate* which assigns quality v(x)ds to each small interval ds spent in state x. The quantity a(x) may be thought of as a state-dependent discount rate. The expected utility of a stochastic tree H which begins at state x and makes transitions to one of the succeeding subtrees K_1, ..., K_n

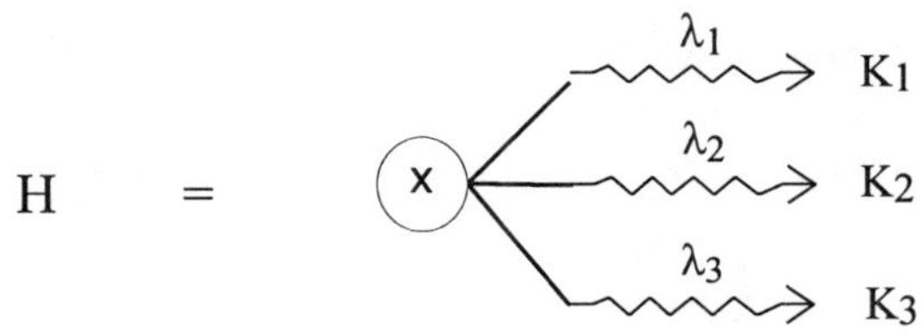

can be shown to satisfy the following recursive equation:

$$E[u(H)] = \frac{v(x) + \sum_i \lambda_i E[u(K_i)]}{a(x) + \sum_i \lambda_i} \tag{3}$$

provided $a(x) + \sum_i \lambda_i > 0$ (otherwise E[u(H)] equals $+\infty$). In this context, the quantity a(x) may be interpreted as the coefficient of constant risk aversion for durations spent in state x. In the risk-neutral case where a(x) = 0, expected utility is equivalent to mean QALY with quality rates v(x). If a(x) is positive, then the decision maker is risk averse for uncertain lifetimes spent in state x, and if a(x) is negative, the decision maker is risk seeking. Note that risk attitude may depend on the health state x. Because expected utility is infinite when a(x) becomes too negative, only limited degrees of risk-seeking preference can be modeled. For practically occurring degrees of risk seeking, we have not found this to be a difficulty.

5. Continuous-Risk Utility Assessment

Continuous-risk utility assessment differs from standard assessment protocols by offering choices between alternatives in which risks are present in continuous time. The assessment scenarios used in standard gamble approaches are simple but they would be too artificial and unrealistic to use in continuous-risk utility assessment. The scenarios we use are realistic and familiar, though more complex.

To see how continuous-risk utility assessment works, consider the scenario of Figure 3, used to assess a subject's risk attitude about future time spent in an incapacitated state (the American College of Rheumatology functional class IV for osteoarthritis of the hip).

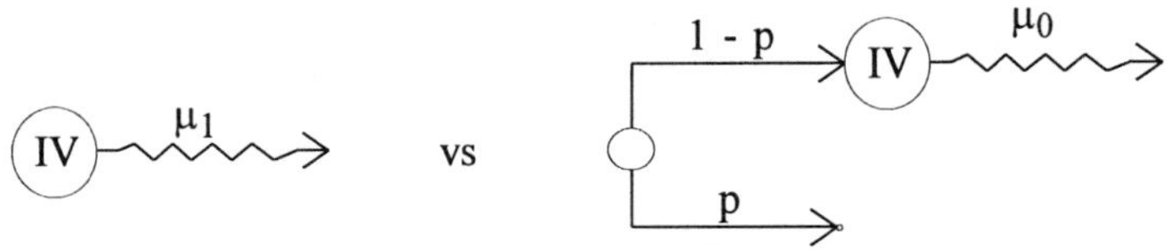

Figure 3: Stochastic tree representation of a continuous-risk assessment scenario

Here the subject is asked to specify what chance p of immediate mortality she is willing to incur in order to decrease her mortality rate from μ_1 to μ_0. Invoking equation (3), a value of p for which indifference holds results in the equality of expected utilities

$$\frac{v(IV)}{a(IV)+\mu_1} = (1-p)\frac{v(IV)}{a(IV)+\mu_0}$$

which determines a(IV):

$$a(IV) = \frac{(1-p)\cdot\mu_1 - \mu_0}{p}.$$

The choices just described are presented to subjects using graphs of exponential survival curves. For example, when $\mu_0 = 0.04386$/yr, $\mu_1 = 2\mu_0 = 0.08772$/yr, and p = 0.18, the choice is between the survival curves S_0 and S_1:

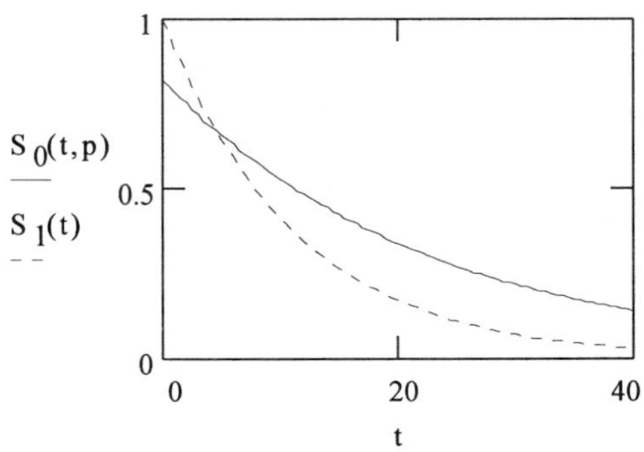

Indifference between the two survival curves would indicate that the subject is willing to take an 18% chance at immediate death in order to cut her ongoing mortality rate in half. The risk-neutral response (producing a(IV) = 0 and giving the same mean lifetimes) is p = p_0 = 0.50. This hypothetical subject is unwilling to take so much immediate risk.

Pellissier and Hazen (1994) performed continuous-risk assessments using an interactive computer program written in TURBO PASCAL. It featured graphical and tabular displays of the survival curves corresponding to both options. Subjects were able to interactively manipulate the probability p of immediate mortality and view the corresponding changes in survival curve, until an indifference value of p was reached. Subjects do successfully respond to the seemingly more challenging continuous-risk questions, and in fact react favorably to the aspects of realism and familiarity present in such scenarios.

6. Risk Attitude and Treatment Choice

We illustrate continuous-risk utility assessment and stochastic tree rollback through the following hypothetical and simplified version of the total hip replacement (Total Hip Arthroplasty or THA) decision. The full model for THA is quite large, and it is presented and analyzed in Hazen, Pellissier and Sounderpandian (1996).

In this model, the patient may be in any one of the four disability classes defined by the American College of Rheumatology. These classes are denoted by I (healthy), II (limited discomfort), III (limited capabilities) and IV (disabled), and in this simplified model, class II is not used. Consider the case of a 60-year-old white female who assigns quality rates v(I) = 1, v(III) = 0.92 and v(IV) = 0.3 to the functional classes I, III and IV. At the time of decision she occupies functional class III. Without surgery, she will progress to status IV at the rate λ = 0.0249/yr. With surgery, she will be returned permanently to functional class I (healthy), unless there are fatal surgical complications that occur with probability 0.25. Her background mortality rate is μ_0 = 0.04386/yr. Stochastic tree representations of the two options are given in Figure 4.

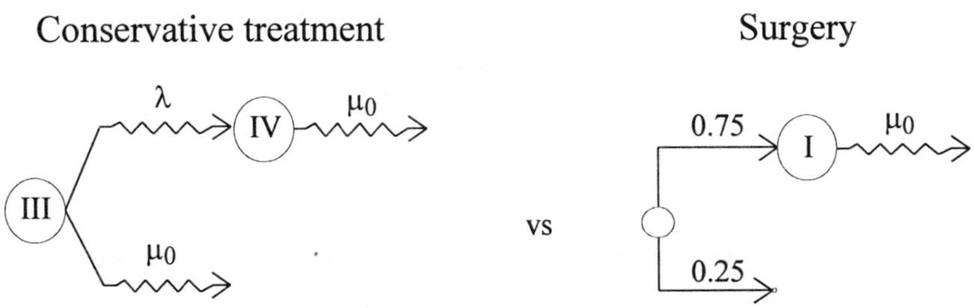

Figure 4: A hypothetical choice between conservative treatment and surgery

Should the woman be risk neutral in all states, that is, a(I) = a(III) = a(IV) = 0, then the rollback formula (2) can be applied to yield overall expected utilities (equal to mean quality-adjusted durations, due to risk neutrality). The results are

$$QALY_{conserv} = 15.857 \text{ yr.} \quad QALY_{surg} = 17.100 \text{ yr.}$$

giving a year-and-a-quarter advantage to the surgical option. On the other hand, should the woman be risk averse with identical coefficients of risk aversion a(I) = a(III) = a(IV) = 0.156 for all states, then rollback gives

$$U_{conserv} = 0.8513 \qquad QRALY_{conserv} = CE_{conserv} = 7.002 \text{ yr.}$$

$$U_{surg} = 0.7500 \qquad QRALY_{surg} = CE_{surg} = 5.645 \text{ yr.}$$

Here $CE_{conserv}$ and CE_{surg} are *certainty equivalent lifetimes* in the healthy class I. For example, the utility of 7.002 years spent in class I is equal to 0.8513, the expected utility of the conservative strategy. Therefore the QRALY or certainty equivalent lifetime of the conservative strategy is $CE_{conserv}$ = 7.002 years. This is a 1.4 year advantage over the 5.645 year certainty equivalent lifetime of the surgical strategy. Note that QRALY and mean QALY coincide under risk neutrality. In this hypothetical example, the change from risk neutrality to risk aversion results in a switch from the more risky surgical strategy to the safer conservative strategy.

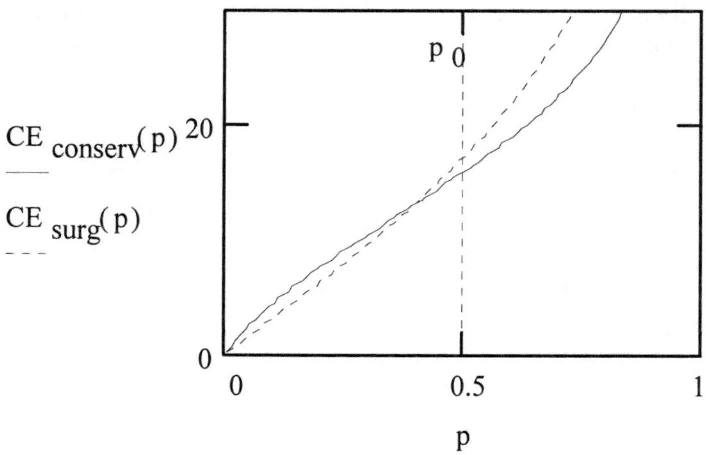

Figure 5: Certainty-equivalent lifetimes Vs hypothetical immediate mortality probability

The values a(x) = 0.156 for x = I, III, IV correspond to an indifference probability p = 0.18 in the continuous-risk assessment scenario of Figure 3. The effect of risk attitude on treatment decision can be captured by graphing the certainty-equivalent lifetimes of the competing strategies against the assessed immediate mortality probability p, as is done in Figure 5. A value of p = 0.4 would result in a tossup between the conservative and surgical treatments.

7. Cost-Effectiveness Analysis

Stochastic tree models can be used to carry out cost-effectiveness analyses. Cost rollback of the trees is performed by assigning time-based cost values, such as for medications and home care, and treating surgical and immediate rehabilitation costs as *tolls* (see Hazen 1992, p.176 for an explanation of tolls). In the analysis of THA below, cost figures were obtained from Northwestern Memorial Hospital data.

Our analysis (see Hazen, Pellissier and Sounderpandian 1996 for details) shows that THA is relatively cost-effective with $10,599/QALY as seen in Table 1. Furthermore, if custodial care costs are also considered, there are many instances when THA may be cost-beneficial. The cost-effectiveness of THA is similar to or better than that of coronary artery bypass surgery ($12,000/QALY to treat symptomatic 3 vessel disease reported in Weinstein and Stason 1985) and that of renal dialysis ($40,000/QALY reported in Roberts et al. 1980), two widely accepted technologies that extend life. Thus, in any policy decision utilizing cost-effectiveness as a criterion, THA should be considered as more effective than these procedures as a means of adding to one's QALY. It may be noted that THA improves QALY by improving the quality rather than the length of life.

TABLE 1. Cost-effectiveness (CE) results for THA.

			Without home care costs		With home care costs
	Life Expectancy	**Mean QALY**	**Expected Cost**	**Marginal CE Ratio**	**Expected Cost**
No THA	22.62 yr.	10.05 yr.	$10,175	-	$198,457
THA	22.25 yr.	18.38 yr.	$98,464	$10,599/yr.	$137,918

8. Concluding Remarks

In this paper, we have introduced the stochastic tree and described how it can be used to perform QALY or QRALY calculations and cost-effectiveness analyses. We have found the stochastic tree applicable for a few other calculations such as the expected time spent in any one state. Overall, we find the application of stochastic tree technique efficient and versatile. The computations required in all these applications can be conveniently done on spreadsheets.

9. References

Beck, J.R. and S.G. Pauker. 1981. Anticoagulation and atrial fibrillation in the brachycardia-tachycardia syndrome. *Medical Decision Making,* 1, pp. 285-301.

Chang, R.W., J.M. Pellissier, and G.B. Hazen. 1993. A cost-effective analysis of total hip arthroplasty for osteoarthritis of the hip. Abstract in *Clinical Research,* 41, 235A, National Meeting, Washington D.C.

Hazen, G.B. 1992. Stochastic Trees: A new technique for temporal medical decision modeling. *Medical Decision Making,* 12, pp. 163-178.

Hazen G.B. 1993. Factored stochastic trees: A tool for solving complex temporal medical decision making problems. *Medical Decision Making,* 13, pp. 227-236.

Hazen, G.B., and J.M. Pellissier. 1994. Recursive utility for stochastic trees. Working paper, Northwestern University.

Hazen, G.B., J.M. Pellissier, and J. Sounderpandian. 1996. Stochastic tree models in medical decision making. Working paper, Northwestern University.

Hillner, B.E., J.P. Hollenberg, and S.G. Pauker. 1986. Postmenopausal estrogens in prevention of osteoporosis. *American Journal of Medicine,* 80, pp. 1115-1127.

Matchar, D.B., and S.G. Pauker. 1986. Transient ischemic attacks in a man with coronary artery disease: Two strategies neck and neck. *Medical Decision Making,* 6, pp. 239-249.

Mooney, C., A.I. Mushlin, and C.E. Phelps. 1990. Targeting assessments of magnetic resonance imaging in suspected multiple sclerosis. *Medical Decision Making,* 10, pp. 77-94.

Pellissier, J.M., and G.B. Hazen. 1994. "Implementation of Continuous Risk Utility Assessment: The Total Hip Replacement Decision," forthcoming in *Socio-Economic Planning Sciences.*

Plante, D.A., J.F. Piccirillo, and R.A. Sofferman. 1987. Decision analysis of treatment options in pyriform sinus carcinoma. *Medical Decision Making,* 7, pp. 74-83.

Roach, P.J., C. Fleming, M.D. Hagen, and S.G. Pauker. 1988. Prostatic cancer in a patient with asymptomatic HIV infection: Are some lives more equal than others? *Medical Decision Making,* 8, pp. 132-144.

Roberts S.D., D.R. Maxwell, and T.L. Gross. 1980. "Cost-effective care of end-stage real disease: A billion dollar question," *Ann Intern Med,* 92, pp. 243-248.

Weinstein, M.C., and W.B. Stason. 1976. *Hypertension: A Policy Perspective.* Harvard University Press, Cambridge, MA.

Weinstein M.C., and W.B. Stason. 1985. "Cost-effectiveness of interventions to prevent or treat coronary heart disease," *Ann. Rev. Public Health,* 6, pp. 41-63.

MANIPULATION OF EMISSION PERMIT MARKETS

E. ROMSTAD & O. BERGLAND
Department of Economics and Social Sciences
Agricultural University of Norway
PO Box 5033
N-1432 Aas
NORWAY

ABSTRACT

Emission permit markets are often thin. The conditions for emission permit markets to yield efficient outcomes are therefore not automatically satisfied. This paper shows – using a principal-agent framework – how the regulatory agency (the principal) can induce price-taking behavior by the firms (the agents) in emission permit markets.

Key words: Transferable emission permits, market manipulation, resource allocation mechanisms.

1. INTRODUCTION

A common critique regarding transferable emission permits is that an emission permit market can be manipulated unless great care is taken in designing and regulating the market. For example in commenting on the efficiency of tradeable emission permits, Hahn (1989) writes:

> "Naturally, these results are subject to the usual cautions that a competitive market actually must exist for the results to hold true". (Hahn, 1989, p. 96)

The question of manipulation of emission permit markets is therefore an important one. Often such markets do not meet the criteria of competitive markets[1], making them prone to manipulation, i.e. that some firm(s) attempt(s) to influence the

[1] Often emission permit markets work as bubbles, with a limited number of potential traders. Other factors making emission permit markets thin are related to the "up-stream – down-stream" problems, i.e. it matters who one trades with.

R. Nau et al. (eds.), Economic and Environmental Risk and Uncertainty, 75–86.

price of emission permits. Mørch von der Fehr (1993) shows that under oligopolistic settings and an inelastic supply of permits, firms may display a more agressive trading behavior in the emission permit market. This raises the question: can the regulatory agency play some strategy which makes manipulation of emission permit markets less likely or even unlikely?

Consider a firm (or a group of firms) trading in an emission permit market. In the short run there is a fairly fixed ratio between outputs and emissions for most technologies. If some of these outputs have an inelastic demand schedule, one opportunity for rent seeking lies in pushing the emission permit price up. The other firms in the sector will sell some of their permits and reduce production, at least in the short run, if their marginal abatement costs differ from the value of the marginal product, and the firm owners do not expect the real value of tomorrow's market price for emission permits to exceed the current permit price. In this paper we show how the regulatory agency can maintain the integrity of the emission permit market through active market participation. Throughout the paper it is assumed that the problem of firms' compliance with their permitted emission levels is solved, i.e. each firm emits according to its quota.[2]

The next section outlines the strategy played by the regulatory agency to induce price taking behavior of the firms in the emission permit market. In the succeeding section the problem of manipulation is discussed in a general theoretical context. Next a principal-agent model is constructed pertaining to the emission permit market.

2. OUTLINE OF THE REGULATORY AGENCY'S STRATEGY

Assume that society's damage function for emissions is known. Moreover assume that an aggregate emission level, Z^o, is determined by the regulatory agency. Let initial emission permits be assigned to the firms[3], so that $\Sigma_{n \in N} z_n^o = Z^o$, where n indexes the firms and z_n^o is the initial allocation of emission permits to firm n. Assuming that emission permit trades take place, there will be an observable market price for emission permits linked to the aggregate emission level, Z^o. If the market price demanded, $p = \overline{p}(Z^o)$, for permits at the aggregate emission level Z^o is above the inferred value of the marginal damages at this emission level, $MD(Z^o)$, the regulatory agency can sell additional permits, thus reducing the emission permit price. Figure 1 depicts the converse situation where the regulatory agency will buy back permits from the firms.

---

[2] Compliance with emission quotas has been discussed in several papers, including Misiolek and Elder (1989); Malik (1990); Russell (1987, 1990); Russell, Harrington and Vaughan (1986); Romstad and Bergland (1993).

[3] The initial allocation of emission permits involves several intriguing questions (auctioning off or grandfathering quotas, and in the latter case, principles for issuing permits), but for now it suffices to assume that these issues are resolved.

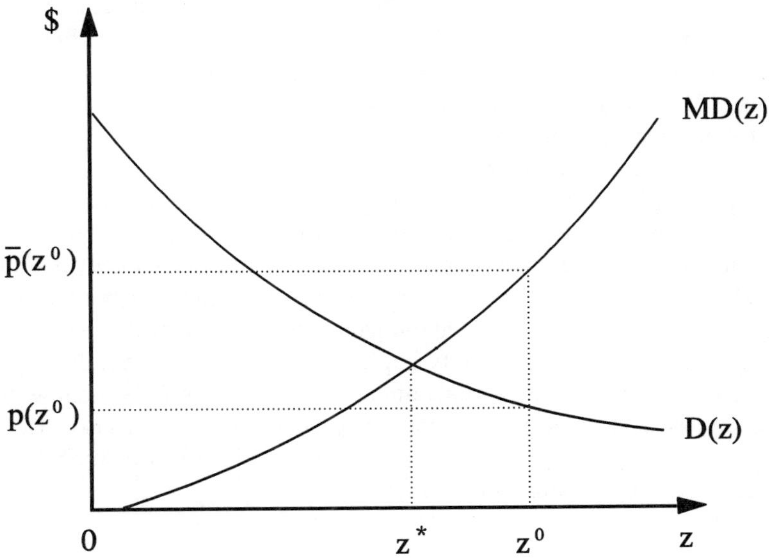

Figure 1: Situation where aggregate emissions (Z^o) are too high.[4]

The intuition behind the proposed mechanism is that even if the emission permit market can be manipulated, it is not in the self interest of any firm to attempt to manipulate the market.

3. ON THE THEORY OF MANIPULATION AND PARETO OPTIMALITY

Unfortunately, achieving both incentive compatibility and Pareto-optimality is not always possible. This is due to the problem of manipulation. The following theorem (Hurwicz, 1972) illustrates the possible impacts of manipulation:

THEOREM 1: Let R′ be a mechanism defined on a family of economic environments, the family of pairs of self-regarding utility functions that exhibit diminishing marginal rates of substitution everywhere. If for every environment within this family of environments, the mechanism R′ generates equilibrium allocations that are Pareto optimal and individually rational, then it can be manipulated.

---

[4] When aggregate emissions are too high resulting in an emission permit price, $p(Z^o)$, that is lower than the inferred price, $\bar{p}(Z^o)$.

The essence of the proof of Theorem 1 is that one agent can obtain increased utility (or profits) by manipulative behavior when all the other agents behave sincerely. Thus each agent is led to behave manipulatively, and the outcome is not Pareto-optimal. Hurwicz (1959, 1973) also shows that in a dynamic market where there is uncertainty about the environment, the dominant strategy of the individual firm is to participate in the market without manipulating the market.

Applying Hurwicz's analysis to emission permit markets indicates that it is in the firms' self interest to participate in a market for emission permits without behaving manipulatively if the aggregate supply of emission permits is not known with certainty. The challenge for the regulatory agency is therefore to find a way of participating in the emission permit market while maintaining the integrity of the market.

It may be difficult to exactly implement such schemes and fully preserve the desired property of welfare maximization. Abreu and Sen (1991) show how almost perfect implementation is possible for a wide class of mechanisms. The suggested mechanism for emission permit markets belongs to this class of problems under the conditions specified by Hurwicz (1973).

4. A PRINCIPAL-AGENT MODEL OF EMISSION PERMIT MARKETS

4.1 THE ECONOMY

The economy consists of firms, consumers and government. Firms produce private goods, while the government orders and pays for public goods.[5] Consumers seek to maximize utility from consumption of both private ($\mathbf{y}$) and public goods ($\mathbf{Z}$). Assume that the objective of the government (through a regulatory agency) is to maximize overall consumer welfare by providing the correct amount of public goods. Let overall consumer welfare be approximated by the following social welfare function:

$$V_t = V_t(M_t, \mathbf{p}_{yt} \mid \mathbf{Z}_t) \tag{1}$$

where:
- V_t denotes a social welfare function indicating societal welfare in monetary terms in time period t,
- M_t is money income in time period t,
- $\mathbf{p}_{yt}$ is a vector of product prices in time period t, and
- $\mathbf{Z}_t$ is a vector of public goods (including emissions).

This social welfare function exhibits the standard properties of indirect utility functions:

[5] Whenever the term "private goods" is used it refers to goods and services that are rival and exclusive, while the term "public goods" refers to non-rival and non-exclusive goods and services (Randall, 1983).

$$\frac{\partial V_t}{\partial M_t} > 0, \text{ and } \frac{\partial V_t}{\partial p_{ytj}} \leq 0 \ \forall \ t \in \mathcal{T}^p, \forall \ j \in J \tag{2}$$

where: $\mathcal{T}^p$ $= \{0, 1, ..., T^p\}$ denotes the index set for the time periods in the agency's planning horizon, and

J $= \{1, 2, ..., J\}$ denotes the index set of produced goods.

For simplicity assume a sector of the economy[6] consisting of one private (y) and one public good (Z), where Z is a public bad like emissions, and that these emissions only stem from this sector. Also assume that the regulatory agency seeks to regulate this sector through tradeable emission permits. The regulatory agency (the principal) must choose a sequence of aggregate emission levels, $\{Z_t\}_{t\in \mathcal{T}^P}$, where $Z_t = \Sigma_{n\in N}\, z_{nt}$, and where $N = \{1, 2, ..., N\}$ denotes the index set of the firms such that social welfare is maximized. The decision problem of the agency then becomes:

$$\underset{\{Z_t\}_{t\in \mathcal{T}^P}}{\text{Max}} \ \Sigma_{t\in \mathcal{T}^P}\, \beta_t \, V_t(M_t, \mathbf{p}_{yt} \mid Z_t) \tag{3}$$

where ß is the discount factor, $(1 + r)^{-1}$.

4.2 THE FIRMS

The firms' cost functions are assumed to satisfy the conditions given in the following cost function existence lemma by McFadden (1978, p. 13). By the duality theorem (McFadden, 1978, p. 82-83) it then follows that the each firm's profit function exists uniquely. The production technology is modelled according to the putty-clay framework of Johansen (1972), i.e. that technology adjustments in the sector do not occur instantly and are costly. Without this assumption there would be few problems assigning quotas in each time period and the firms' optimization problems would degenerate to a single-period decision. Another and even more important implication of the putty-clay assumption is that emission permits need to be awarded permanently (or for several time periods) to avoid too much volatility in the firms' decision environments. The nth firm's profit maximizing problem in a situation with tradeable emission permits can then be expressed the following way:

$$\underset{\{y_{nt}\ \bar{z}_{nt}\}}{\text{MAX}} \ \Sigma_{t\in \mathcal{T}}\, \beta^t \, \{p_{yt}[y_{nt} + \hat{Y}_{-nt}]\, y_{nt} - C_{\theta t}(y_{nt}, \mathbf{v}_t; z_n^o + \bar{z}_{nt}) - p_{zt}(Z_t)\, \bar{z}_{nt}\} \tag{4}$$

where: $\mathcal{T}$ $= \{0, 1, ..., T^n\}$, the index set for the time periods in firm n's planning horizon,

p_{yt} is the market price for the produced good, y_t,

y_{nt} is the nth firm's output in period t,

$\hat{Y}_{-nt}$ is firm n's expectations on the other firms' output in period t,

---

[6] A sector of the economy concerns the production and consumption of a particular type of goods or services.

z^o_n is the (permanent) emission permits awarded to the nth firm in period t=0,

$\bar{z}_{nt}$ is the emission permits bought by the nth firm in period t,

$C_{\theta t}$ is the (to the agency unknown) cost function of firm n of producing y_{nt} with firm n's total emission level, $z_{nt} = z^o_{nt} + \bar{z}_{nt}$,

v_t is a vector of input prices for regular inputs, and

p_{zt} is the price of the emission permits in the emission permit market.

Assume that $\frac{\partial y_{\theta t}}{\partial z_{nt}} > 0$ for any production technology θ, $\theta \in \Theta$, where Θ denotes the index set of available technologies. Then for a given level of emissions, z_{nt}, there exists a single level of output, $y_{\theta t} = y_\theta(z_{nt})$. For the marginal production costs to be non-negative for any emission level, $\frac{dC_{\theta t}}{dz_{nt}} = \frac{\partial C_{\theta t}}{\partial y_{nt}}\frac{\partial y_{nt}}{\partial z_{nt}} + \frac{\partial C_{\theta t}}{\partial z_{nt}} > 0$ must hold, requiring $\left| \frac{\partial C_{\theta t}}{\partial y_{nt}} \cdot \frac{\partial y_{nt}}{\partial z_{nt}} \right| > \left| \frac{\partial C_{\theta t}}{\partial z_{nt}} \right|$. This makes it possible to reformulate (4) into a decision problem with the emission level, z_{nt}, being the only choice variable.

$$\left\{\begin{matrix}\text{MAX}\\ z_{nt}\end{matrix}\right\} \Sigma_{t\in T}\, \beta^t \{p_{yt}[y_{\theta t}(z_{nt}) + \hat{Y}(Z_{-nt})]y_{\theta t}(z_{nt}) \quad\quad (5)$$

$$- C_{\theta t}(v_t;z_{nt}) - p_{zt}(z_{nt}+Z_{-nt})(z_{nt} - z^o_n)\}$$

where: $\hat{Y}(Z_{-nt})$ is firm n's expectations of what the other firms in the sector produce in time period t, and

Z_{-nt} is firm n's assessment of what the other firms in the sector emit in time period t.

For simplicity assume that all firms are risk neutral. The uncertain elements in firm n's multi-period decision problem are the aggregate supply of the other firms in the sector (and hence the product price), and the future prices on emission permits, which again are linked to the sequence of the aggregate supply of permits, $\{Z_t\}_{t\in T}$.

4.3 A PRINCIPAL-AGENT MODEL OF THE EMISSION PERMIT MARKET

The principal-agent formulation is the game between the agency and the firms defined by (3) and (5). Consistent with earlier works on emission controls (see for example Baumol and Oates, 1988; Xepapadeas, 1991), it is assumed that the regulatory agency does not know the exact production technology of individual firms. To analyze the problem of manipulative behavior of firms, the following definitions are needed:

DEFINITION 1: *Manipulative behavior* in the emission permit market for the nth firm is any action different from the action defined by the nth firm's standard cost minimizing demand curve for emission permits.

DEFINITION 2: A *coalition* of firms is a collection of firms coordinating their strategies.

The aim of manipulative behavior of any firm in the emission permit market is to increase its discounted profits. It will be shown that it is not profitable for any firm to attempt to manipulate the emission permit market, alone or in a coalition if the regulatory agency commits to the suggested strategy.

Profit maximization for firm n implies that along the optimal trajectory of emission permits to hold for firm n the following condition holds:

$$\left[\frac{\partial p_{yt}}{\partial y_{\theta t}} y_{\theta t}(z_{nt}) + p_{yt}\right] \frac{\partial y_{\theta t}}{\partial z_{nt}} = -\frac{\partial C_{\theta t}}{\partial z_{nt}} = \frac{\partial p_{zt}}{\partial z_{nt}} z_{nt} + p_{zt} \tag{6}$$

For any firm fully utilizing its arbitrage opportunities between time periods, the following must hold for all time periods $t \in T$:

$$-\frac{\partial C_{\theta t}}{\partial z_{nt}} = \left[\frac{\partial p_{zt}}{\partial z_{nt}} z_{nt} + p_{zt}\right] = -\beta \frac{\partial C_{\theta t+1}}{\partial z_{nt+1}} = \beta \left[\frac{\partial \hat{p}_{zt+1}}{\partial z_{nt+1}} z_{nt+1} + \hat{p}_{zt+1}\right] \tag{7}$$

The practical implication of (7) is that the cost minimizing strategy of firm n is to buy (sell) permits today if it expects the discounted permit price in future periods to be above (below) the current permit price.

There are two parts to the proof that non-manipulative behavior is a dominated strategy. Part one shows that non-manipulative behavior in the emission permit market is the *best reply strategy for a single firm* provided that the other agents are price takers and the regulatory agency commits to its outlined strategy.

The second part of the proof shows that the other firms behave as price takers in the emission permit market. The intuition behind the first part of the proof is that no firm will attempt to push the emission permit price up, if it knows that this will cause the aggregate supply of permits to increase and the subsequent emission permit price to fall.

Figure 2 depicts how a firm attempting to push the market price for emissions up by buying permits at a higher price than necessary, as the regulatory agency may increase the aggregate supply of permits in the next time period. Implicitly this results in the firm not exploiting its arbitrage opportunities, meaning that (7) no longer holds.

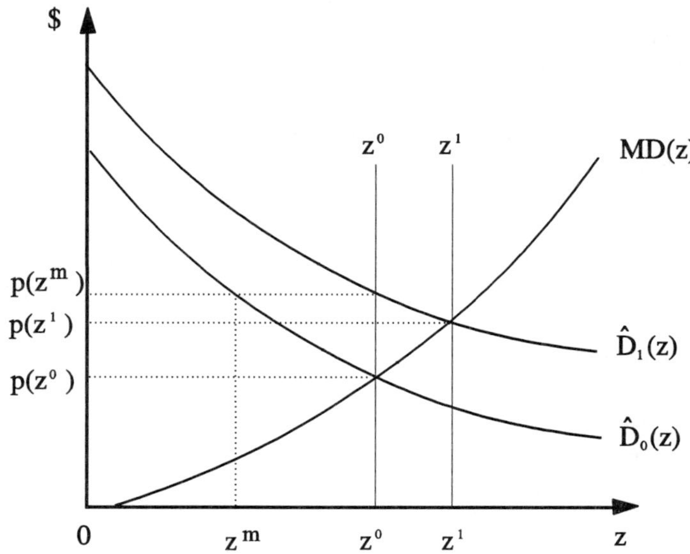

Figure 2: Situation where a single firm attempts to push the market price for emission permits up. [7]

To prove part one the following lemmas are needed:

LEMMA 1: The market price for emission permits, p_{zt}, is reduced with an increase in the amount of emission permits (Spulber, 1989, p. 353).

LEMMA 2: The assumptions leading to definition of the cost function with externalities, $C_{\theta t}(y_{nt},v_t;z_{nt})$, assure that the nth firm's demand curve for emitting z_{nt} is single valued, i.e. for a given output level, y_{nt}, there exists one and only one emission level. As the profit function is the convex conjugate of the cost function, it therefore follows that to the profit maximizing output level, y^*_{nt}, there exists one emission level, z^*_{nt}.

PROOF: Follows by the transition from (4) to (5).

Q.E.D.

[7] $\hat{D}_0$: the regulatory agency's initial estimate of the demand for permits, $\hat{D}_1$: the regulatory agency's revised estimate of the demand for permits after firm n has attempted to push the permit price up by buying additional permits ($Z^o - Z^m$), $p(Z^o)$: initial market price for permits, $p(Z^m)$: market price for permits just before the agency sells additional permits, $p(Z^1)$: market price for permits after additional permits have been sold by the regulatory agency.

PROPOSITION 1: Non-manipulative behavior is the best reply strategy for any single agent provided that the other agents act as price takers in the emission permit market.

PROOF: By Lemma 2 there exists one emission level, z^*_{nt}, to each profit maximizing output level, y^*_{nt}. Suppose one agent attempts to manipulate the permit market by buying emission permits such that his stock of permits, $z'_{nt} > z^*_{nt}$. As (6) no longer holds, the expected profits of the manipulating agent are less than they would be under non-manipulative behavior. By Lemma 1 the market price for permits increases as the overall demand for permits increases. With the increase in the permit price, the regulatory agency increases the aggregate emission quota by offering additional permits for sale, which again lowers the permit price. This causes the manipulative agent to be further off from the necessary trajectory for profit maximization (6). The only reason for any agent to hold a quantity of permits above what is indicated by (6) is that the he expects future permit prices to increase even more. As all agents know that the regulatory agency eventually will increase the aggregate supply of permits, they also know that future permit prices will not increase forever. Hence, some agents will sell their permits, reducing the rate of permit price increase or even cause the permit price to fall. Consequently the expected discounted profits from manipulative behavior are less than the expected discounted profits from non-manipulative behavior.

Q.E.D.

The intuition behind the second part of the proof is that in markets with uncertainty, expected profit maximizing agents will eventually sell (buy) to safeguard against losses.

PROPOSITION 2: As the future aggregate supply of permits will change with probability one at some random future time period, agents will trade permits when the market price for emission permits deviates from their marginal abatement costs and to the extent that all expected arbitrage opportunities are exploited.

PROOF: Suppose an agent observes increasing (declining) prices in the permit market for some periods. The agents decision problem is then to choose an action that minimizes his/her Bayes' risk. Agents, knowing with certainty that under prolonged increasing (declining) prices, the regulatory agency will eventually sell (buy) permits to make the emission permit price equal to the marginal damages from emissions, the strategy that minimizes the agent's Bayes' risk is to sell (buy) permits. By backwards recursion it then follows that the agent will sell (buy) permits so that (6) and (7) holds. This is observationally equivalent with the agent being a price taker in the emission permit market.

Q.E.D.

By playing the suggested strategy in the emission permit market, the regulatory agency induces non-manipulative behavior from a single firm provided that the other firms behave as price takers in the emission permit market. However the outlined strategy is potentially vulnerable to manipulation by a sufficiently large coalition of firms. To see this assume that a group of firms jointly started buying permits to raise the permit price. Once the permit price exceeds the inferred value at the given aggregate emission level, the agency would then sell additional permits, increasing the aggregate supply of permits (and hence emissions).

Provided that the emission permit markets is set up like a stock exchange, the anonymity of firms trading in the market with respect to other firms is preserved. This makes it difficult for the colluding firms "sticking to their agreement" to sanction firms breaking the agreement. Hence such a coalition of firms is unstable, and efforts of joint manipulation by several firms is unlikely.

DEFINITION 3: A *stable coalition* is a coalition where the expected profits of every firm in the coalition exceeds the expected profits of belonging to any other coalition or belonging to no coalition (Cornes and Sandler, 1986, p. 96).

PROPOSITION 3: Provided that the anonymity of firms trading in the emission permit markets is preserved with respect to other firms, any coalition of manipulative firms in the permit market is unstable as any member of the coalition has greater expected profits from non-manipulative behavior.

PROOF: Follows directly from Proposition 1 and 2. As the expected profits from manipulative behavior is lower than the profits from non-manipulative behavior in the short run, and colluding firms have limited opportunities to sanction firms breaking the collusive agreement, any member of the coalition has incentives to pursue the profit maximizing behavior consistent with (6). Hence the coalition is unstable.

Q.E.D.

5. CONCLUDING REMARKS

Contrary to the findings of amongst others Mørch von der Fehr (1993), this paper has demonstrated how the regulatory agency can induce non-manipulative behavior in the emission permit market. This is achieved by introducing uncertainty in an emission permit market (confer with Hurwicz, 1973). This uncertainty is achieved by committing to a strategy of adjusting the aggregate supply of permits through buying (selling) if the market price for emission permits is below (above) the value obtained from the marginal damage curve for emissions, evaluated at the aggregate emission level. Despite the fact that the suggested strategy of the regulatory agency is vulnerable to manipulation by a group of firms, such a joint effort to manipulate the emission permit market is unlikely as any such coalition is unstable. Using the insights of approximate implementation (Abreu and Sen, 1991) the resulting market solution comes close to maintaining the desired efficiency of the emission permit market. This has important implications for the applicability of emission permit markets to induce optimal emission levels when firms are reluctant to disclose their abatement costs (Romstad and Bergland, 1990).

As noted earlier, firms are assumed to have putty-clay technologies, i.e. that technology changes do not happen instantly and are costly. Therefore the initial permits to firms must be awarded on a permanent basis (or at least for more than one time period). This implies that the traded emission permits must be viewed as a capital asset. Consequently the regulatory agency cannot compare the price of emission permits with the inferred value. Thus the appropriate comparison with the inferred value is the capital costs (price times the interest rate) of acquiring the permits.

Other important aspects of emission permit markets include the problem of non-participation and transaction costs (see Roberts, 1982; Hahn and Noll, 1982; Hahn, 1983 and 1989; Atkinson and Tietenberg, 1991). By constructing the emission permit markets like a stock exchange, prices are observable by all agents at little costs, transaction costs are kept to a minimum, and the anonymity of firms trading in the market is preserved with respect to other firms. The latter is important to ensure the instability of any coalition of firms attempting to manipulate the emission permit market.

BIBLIOGRAPHY

Abreu, D. & A. Sen (1991): "Virtual Implementation in Nash Equilibrium", *Econometrica*, **59**(4):997-1021.

Atkinson, S. & T. Tietenberg (1991): "Market Failure in Incentive-Based Regulation: The Case of Emissions Trading", *Journal of Environmental Economics and Management*, **21**(1):17-31.

Hahn, R.W. (1989): "Economic Prescriptions for Environmental Problems", *Journal of Economic Perspectives*, **3**(2):95-114.

Hahn, R.W. & R.G. Noll (1982): "Designing a Market for Tradeable Emissions Permits", pp. 119-146 in Magat, W.A. (ed.)*Reform of Environmental Regulation*, Ballinger Publishing Company, Cambridge, MA.

Hurwicz, L. (1959): "Optimality and Informational Efficiency in Resource Allocation Processes", pp. 27-46 in Arrow, K.J., S. Karlin & P. Suppes (eds.) *Mathematical Methods in the Social Sciences*, Stanford University Press, Paolo Alto, CA.

Hurwicz, L. (1972): "On Informationally Decentralized Systems", pp. 297-336 in McGuire, C.B. & R. Radner (eds.) *Decisions and Organizations*, North-Holland, Amsterdam, The Netherlands.

Hurwicz, L. (1973): The Design of Mechanisms for Resource Allocation", *American Economic Review*, **63**(1)1-30.

Johansen, L. (1972): *Production Functions*, North-Holland Publishing Company, Amsterdam, The Netherlands.

Malik, A.S. (1990): "Markets for Pollution Control When Firms are Noncompliant", *Journal of Environmental Economics and Management*, **18**(2):97-106.

McFadden, D. (1978): "Duality of Production, Cost, and Production Functions", pp. 1-109 in Fuss, M. & D. McFadden (eds.) *Production Economics: A Dual Approach to Theory and Applications*, Vol. 1., North-Holland, Amsterdam, The Netherlands.

Misiolek, W.S. & H.W. Elder (1989): "Exclusionary Manipulation of Markets for Pollution Rights", *Journal of Environmental Economics and Management*, **16**:156-166.

Mørch von der Fehr, N-H (1993): "Tradeable Emission Rights and Strategic Interaction", *Environmental and Resource Economics*, **3**:129-151.

Randall, A. (1983): "The Problem of Market Failure", *Natural Resources Journal*, **23**(1):131-148.

Roberts, M.J. (1982): "Some Problems of Implementing Marketable Pollution Rights Schemes: The Case of the Clean Air Act", pp. 92-118 in Magat, W.A. (ed.) *Reform of Environmental Regulation*, Ballinger Publishing Company, Cambridge, MA.

Romstad, E. & O. Bergland (1990): *Principles for Obtaining Optimal Emission Levels When Abatement Costs Are Private Information*, Oregon State Graduate Faculty Working Paper No. 90-103.

Romstad, E. & O. Bergland (1993): "Inducing Individual Firm Compliance to Emission Quotas When Abatement Costs are Private Knowledge", paper presented at the Fourth EAERE Conference, Fontainebleu, France, June 30 - July 3, 1993.

Russell, C.S. (1987): "Game Theory Lessons for Structuring Monitoring and Enforcement Systems", paper presented at the Association of Environmental and Resource Economists' Workshop on Environmental Monitoring and Enforcement: Theory and Practice, Newark, Delaware, August 13-14, 1987.

Russell, C.S. (1990): "Game Models for Structuring Monitoring and Enforcement Systems", *Natural Resources Modelling*, **4**(2):143-173.

Russell, C.S., W. Harrington & W. J. Vaughan (1986): *Enforcing Pollution Control Laws*, Resources for the Future, Washington D.C.

Spulber, D.F. (1989): *Regulation and Markets*, MIT Press, Cambridge, MA.

Xepapadeas, A.P. (1991): "Environmental Policy under Imperfect Information: Incentives and Moral Hazard", *Journal of Environmental Economics and Management*, **20**(2)113-126.

II. Risk and Uncertainty in Economic Theory

ALLAIS THEORY OFFERS EXPLANATION FOR EQUITY PREMIUM PUZZLE[1]

JOHN E. GOLOB
Federal Reserve Bank of Kansas City
925 Grand Boulevard, Kansas City, Missouri 64198 USA

1. Introduction

Dynamic equilibrium models of the macroeconomy are often unable to replicate the asset returns observed in financial markets. Both Lucas [1978] endowment economies and equilibrium business cycle models typically predict either lower risk premiums or higher risk-free rates than those found empirically. The low risk premium was designated the "equity premium puzzle" by Mehra and Prescott [1985], and this puzzle has stimulated extensive research into alternative specifications for dynamic equilibrium models[2]. This paper shows that a model with nonexpected utility preferences can replicate the asset returns observed in financial markets.

The paper calculates asset returns in a Lucas endowment economy with "linearized moments" preferences, which are based on the Allais [1952] theory of risk. Mùnera and de Neufville [1983] have shown that these preferences can successfully predict choices over static gambles, including Allais paradox experiments that are inconsistent with expected utility. The linearized moments approach is generalized to a dynamic setting by constructing a discounted time-separable extension of the static model. With

[1]The author is grateful to Andrew Abel, Franklin Allen, Marshall Blume, Colin Camerer, Bob Chirinko, Frank Diebold, Gary Gorton, Craig Hakkio, Richard Kihlstrom, Guy Laroque, Howard Kunreuther, Raj Mehra, Robert Pollak, Zvi Safra, Anne Sibert, Oren Sussman, Patrick Waldron, John Weinberg, and Steve Zeldes for comments on earlier versions of this paper. The author also acknowledges helpful discussions with workshop participants at the Federal Reserve Bank of Kansas City, the University of Michigan, and the Federal Reserve System Committee on Financial Analysis. The views expressed herein are solely those of the author and do not necessarily reflect the views of the Federal Reserve Bank of Kansas City or the Federal Reserve System.

[2]See Abel [1991] and Kocherlakota [1995] for surveys on the equity premium puzzle.

R. Nau et al. (eds.), Economic and Environmental Risk and Uncertainty, 89–108.

risk parameter estimates from static choice experiments, the dynamic model is consistent with the risk premium and risk-free rate observed in financial markets.

The Allais theory of risk contains two elements. First, a cardinal utility function represents "intensity of preferences" and is independent of risk. Second, risk attitudes are specified by higher moments of uncertain utility. In contrast, expected utility theory considers only the first moment of uncertain utility and characterizes risk aversion by the curvature of the utility function. From Allais's perspective, expected utility is inadequate because the shape of the utility function simultaneously specifies two attributes of preferences, risk and intensity of preferences. Section 2 shows how the equity premium puzzle can be explained as a failure to distinguish between these attributes.

In addition to Allais's critique of expected utility for evaluating static gambles, expected utility has a more widely recognized limitation in dynamic models. The limitation is that the curvature of the utility function simultaneously determines both risk aversion and intertemporal substitution. Epstein and Zin [1989] and Weil [1990] have shown that risk aversion can be separated from intertemporal substitution with a class of recursive preferences introduced by Kreps and Porteus [1978]. The linearized moments approach also separates risk from intertemporal substitution, but with a time-separable rather than a recursive structure.[3]

The paper is structured as follows: Section 2 provides a brief description of the Allais theory of risk and shows how the Allais critique of expected utility provides an explanation for the equity premium puzzle. Section 3 introduces linearized moments preferences as a special case of the Allais theory and discusses estimates of the preference parameters. Section 4 shows that the risk premium and risk-free rate in the Lucas asset pricing model with linearized moments preferences are consistent with returns observed in financial markets. Section 5 concludes.

2. Explaining the Equity Premium Puzzle

This section offers an explanation for the equity premium puzzle from the perspective of the Allais critique of expected utility. The behavior of the risk-free rate is central in this explanation. Weil [1989] coined the term "risk-free rate puzzle" to describe the behavior of the risk-free rate in dynamic models. He noted that his model would yield a substantial risk pre-

[3] An important feature of the Kreps-Porteus approach is that consumers are not indifferent about the time when uncertainty is resolved, which is not a feature of the model presented in this paper. Kreps-Porteus preferences have not been successful, however, at resolving the equity premium puzzle (Weil, 1989).

mium only if the risk-free rate was much larger than the rate observed in financial markets. This section explains why dynamic models can generate anomalously high risk-free rates.

2.1. THE ALLAIS THEORY

Allais's theory contains two basic elements: (1) the existence of a cardinal utility function that is independent of risk attitudes and (2) the valuation of risky lotteries depends on multiple moments of the probability distribution of uncertain utility.

The first element of the Allais theory is a cardinal utility function. Although the function in the expected utility theory has also been described as cardinal, Allais's utility function is cardinal in a different way.[4] That is, Allais's utility function measures the intensity of preferences, as in the view of measurable utility that was common among nineteenth-century neoclassical economists. In contrast, the cardinality in expected utility is associated with risk attitudes.

Both the Allais theory and the expected utility theory use a function to transform uncertain outcomes into uncertain utility. When the argument of the utility function is consumption, as in this paper, the function maps a probability distribution defined over consumption into a probability distribution defined over utility. The theories differ, however, in the number of moments that are relevant. Only the first moment of uncertain utility, the expected value, is relevant in expected utility theory. In contrast, risk attitudes are determined by the higher moments of the probability distribution of utility in the Allais theory.

Allais's criticism of expected utility is based on his view that utility is measurable under certainty, i.e. is cardinal. From this view, expected utility theory combines two independent attributes of preferences in the utility function, risk and intensity of preferences. The equity premium puzzle can be traced to expected utility's characterization of risk attitudes with curvature of the utility function.

2.2. THE RISK-FREE RATE

While the anomalous behavior of the equity premium and the risk-free rate extends to many dynamic models, the discussion in this paper focuses on the specific model examined by Mehra and Prescott [1985]. They consider a Lucas [1978] "fruit tree" economy, in which all income is paid as dividends of perishable consumption goods to the economy's asset holders. Equilibrium

[4] Baumol [1958] offers an extensive discussion of the distinction between cardinal utility in the neoclassical sense versus the utility index in expected utility theory. Also see Fishburn [1988].

asset prices are determined by a representative agent maximizing expected utility over an infinite horizon subject to a budget constraint

$$V_t = \max_{a_t} E_t \sum_{j=0}^{\infty} \beta^j u(c_{t+j}), \quad s.t. \quad c_t = a_{t-1}(p_t + y_t) - a_t p_t, \tag{1}$$

where β is the time discount factor and c_t, p_t, y_t, and a_t are consumption, asset prices, dividends, and asset holdings respectively at time t. The form of the utility function is constant relative risk aversion (CRRA), $u(c) = c^{1-\alpha}/(1-\alpha)$, where α is the coefficient of risk aversion. Dividend growth is assumed to follow a Markov process calibrated to U.S. per capita consumption data, which in the two-state version of the model grows 5.4% annually during an expansion and contracts by 1.8% annually during a recession. Under the parameter restrictions imposed by Mehra and Prescott, the maximum excess return of the risky over the risk-free asset was only 0.35%, compared to 6% in historical data. Although this is a relatively simple dynamic model, generalizations that include labor income and investment typically reduce the risk premium even further, because the return to the risky asset will tend to be greatest in the simple case when dividends are perfectly correlated with consumption.

Figure 1 shows how the returns to risky and risk-free assets in the Mehra-Prescott model depend on the risk aversion parameter α in an economy with i.i.d. growth.[5] Although the risky rate increases with α, the risk-free rate also increases. For example, while the risky return is 15 percent when $\alpha = 10$ (the maximum allowed by Mehra-Prescott), the risk-free rate is 14 percent, yielding only a 1 percent risk premium. Since the equity premium puzzle can be attributed to the behavior of the risk-free rate, the next question is: why does the risk-free rate increase with the risk aversion parameter?

The behavior of the risk-free rate in dynamic equilibrium models can be explained by examining a general expression for the price of an asset

$$p_t = E_t \sum_{j=1}^{\infty} \beta^j \frac{u'(c_{t+j})}{u'(c_t)} y_{t+j}, \tag{2}$$

where the symbols are defined as in equation (1). In consumption models of asset pricing, the dividends y_{t+j} are always multiplied by the marginal utility of consumption $u'(c_{t+j})$. That is, in a dynamic model the "value" of an asset's dividend is determined by its marginal utility. This implies that a risk-free asset can be implicitly risky. Even if the payoff is known, the marginal utility of the payoff may be highly variable.

[5]The details of solving the model are contained in the article by Mehra and Prescott.

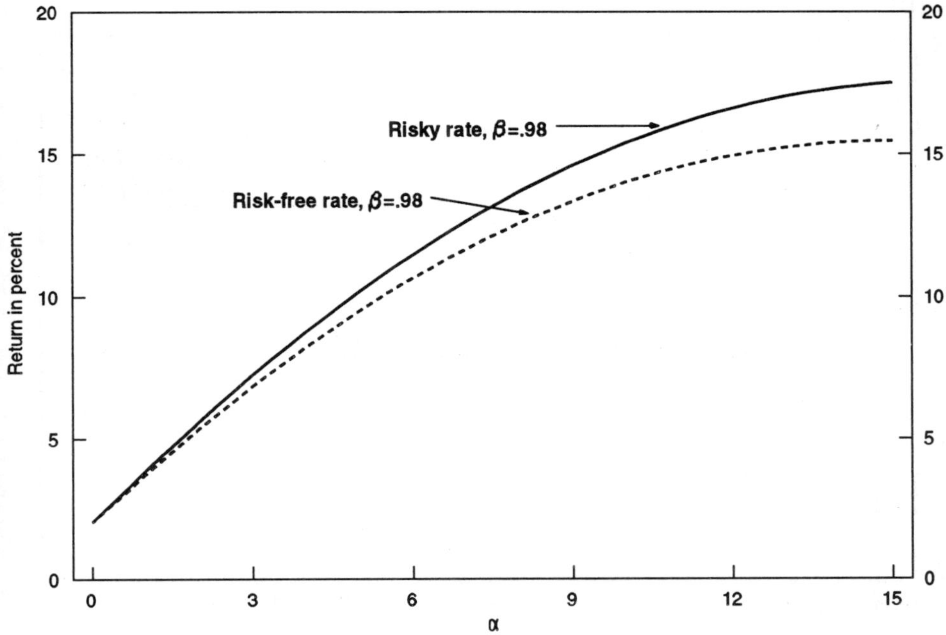

Figure 1. Asset Returns vs. Risk Aversion parameter (α)

Figure 2 illustrates the variability of marginal utility over a one-year horizon for a CRRA utility function with $\alpha = 10$ (the curve is normalized to unity at $c = 1$).[6] Figure 2 includes tangent lines for the $u(c)$ curve at $c = 1.054$ and $c = 0.982$, to compare the marginal utilities at time $t + 1$ for an economic expansion and contraction in a two-state Markov economy. For $\alpha = 10$ the marginal utility is twice as high during the recession as the expansion, so the "value" of the risk-free asset is twice as high during the recession. Thus, because the value of the dividend paid to the risk-free asset is variable the asset is implicitly risky, and therefore, its return is high.

Figure 2 illustrates that when risk aversion in the model is large, the variability of marginal utility becomes the dominant source of risk. The dividends of the risky asset vary by only a few percent, much less than the factor of two variation in marginal utility. Thus, when asset prices are determined by the marginal utility of their dividends, the risky and risk-free assets are almost equally risky.

[6]While the example is based on the CRRA utility function, the discussion applies equally well to other functional forms, which can be approximated locally by a CRRA function.

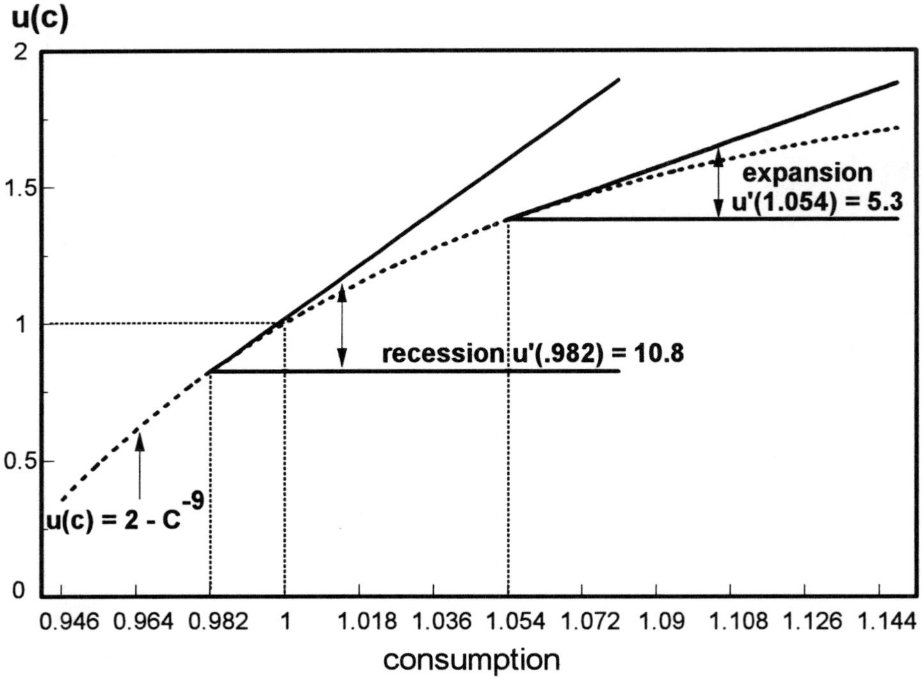

Figure 2. Variation in Marginal Utility for CRRA Functions

Having established that the equity premium puzzle can be attributed to the behavior of the risk-free rate, it is time to return to the Allais view of risk. In the Allais theory, risk aversion is separate from the shape of the utility function. Thus, the problem with expected utility is that high risk aversion requires high curvature in the utility function, which causes the marginal utility of the "risk-free" payoff to be uncertain, so even the risk-free asset is implicitly risky. In contrast, the Allais cardinal utility function has low curvature, so the marginal utility of the risk-free payoff has minimal variability and the asset is truly risk-free.[7]

3. The Linearized Moments Model

If the cause of the equity premium puzzle is the limitation of expected utility described in Section 2, a model based on the Allais theory should yield asset returns that are consistent with historical data. Before exploring this question, the Allais theory be must be generalized to a dynamic model. Previous research with the Allais theory has focused on static choice ex-

[7] Allais's [1988] empirical estimates confirm that the cardinal utility function has little curvature.

periments, such as the Allais paradox, whereas the equity premium puzzle is a feature of dynamic models. This section describes a static model and Section 4 develops a dynamic extension.

Mùnera and de Neufville [1983] have analyzed a model for risky decisions that is based on "linearized moments."[8] Munier [1989, p. 312] describes the model as a Taylor approximation to a more general Allais functional. Since the linearized moments model uses both a cardinal utility function and higher moments of the probability distribution of utility, it is consistent with the Allais theory. Mùnera and de Neufville have shown that the model is consistent with empirical preferences over static gambles that are inconsistent with expected utility. The model is useful for the analysis in this paper because Mùnera [1988] has estimated the model's parameters from preferences expressed by experimental subjects. To the extent that microeconomic foundations are viewed as important in dynamic macroeconomic models, the parameters should be consistent with micro data.

The cardinal utility function in the linearized moments model will be denoted by $u(\cdot)$. For a lottery X with uncertain outcomes x_k with probability p_k, $k = 1, \ldots, n$ and $p_1 + p_2 + \cdots + p_n = 1$, the preference functional $U(X)$ is defined as the expected value of $u(X)$ plus a linear function of the higher linearized moments of $u(X)$

$$U(X) = E[u(X)] + \sum_{k=2}^{r} \gamma_k M_k \tag{3}$$

$$\text{where} \quad M_k = \left\{E\left[u(X) - u(x_o)\right]^k\right\}^{1/k} \quad , \quad k \geq 2.$$

$E(\cdot)$ is the expectation operator, $u(x_0)$ is a reference level that will be chosen to be the expected value of $u(X)$, and r is the number of moments in the model. The M_k's are referred to as linearized moments because for a constant b, $M_k(bu(\cdot)) = bM_k(u(\cdot))$. In Mùnera's terminology the γ_k parameters capture attitudes-toward-randomness, which is related to the concept of risk aversion in expected utility.

Specifying the model requires estimates of the γ_k parameters. Unfortunately, the structure of the model does not allow a generalized method of moments (GMM) estimate of these parameters.[9] Nevertheless, Mùnera

[8] See Schoemaker [1982], Weber and Camerer [1987], and Munier [1989] for general discussions of alternative nonexpected utility models of behavior under risk.

[9] GMM estimates require converting a conditional expectation into an unconditional expectation. Dynamic expected utility models typically have a first-order condition of the form $E_t f(x_t, \theta_t) = 0$, where E_t is the expectation operator and $f(.,.)$ is a function of a vector of observable variables x_t and parameters θ_t. By applying the law of iterated expectations, the condition expectation becomes an unconditional expectation. But in a model with higher moments, the first-order condition is of the form $E_t f(x_t, E_t x_t, \theta_t) =$

[1988] has estimated parameters by fitting a three moment version of the model to experiments conducted by other researchers. He estimated different versions of the model for four categories of static choice experiments: (1) loss of life, (2) saving life, (3) loss of money, and (4) winning money. Assuming that investor behavior is best captured by the latter category, I take these parameter values in the analysis that follows.[10] Although the variation of parameter estimates across different versions of the model reduces its generality, if the model is viewed as a Taylor approximation to a more general function of the moments, it is not surprising that the parameters vary across different categories of experiments.

For the model written in equation (3) the parameter estimates from Mùnera are in the ranges $0.21 < -\gamma_2 < 1.1$ and $0.4 < \gamma_3 < 4.2$.[11] Note that the signs of the coefficients are consistent with generally accepted attitudes towards risk, with negative γ_2 implying less utility for gambles with greater standard deviation. With respect to the third moment, the positive coefficient is in agreement with Hirshleifer [1970], who has noted a preference for positive skewness among investors. The positive sign for γ_3 is also consistent with Friedman and Savage's [1948] observation that individuals simultaneously buy lottery tickets and insurance.[12] That is, the large M_3 of a lottery makes a positive contribution to utility, while insurance protects against the negative M_3 of a low-probability disaster. Friedman and Savage explained this behavior in the context of expected utility by postulating a utility function with two concave segments separated by a convex segment. But Markowitz [1952] noted that this explanation is not consistent with the observation that individuals at all levels of wealth buy both lottery tickets and insurance. In the linearized moments model, a positive coefficient on M_3 can be consistent with the purchase of both lottery tickets and insurance over all levels of wealth.

To see whether Mùnera's parameter estimates are reasonable, the model is examined in the context of Allais paradox experiments. Table 1 shows two versions of the Allais paradox, the first from Kahneman and Tversky [1979] and the second from Hagen [1988]. Each version of the paradox consists of four gambles, and the table shows the probability of different dollar out-

0. The unconditional expectation of this expression is $Ef(x_t, E_t x_t, \theta_t) = 0$, and the conditional expectation within the function precludes a GMM estimate.

[10]Kahneman and Tversky [1979] have shown a preference for more randomness in experiments where subjects expect to lose money. This result is not consistent with the usual assumption in economic and financial models that agents prefer less randomness.

[11]The parameters are taken from Mùnera [1988] page 302, Table II, Model 2 of experiments 1 and 3. This is the model that offers the best fit for the experiments in which money is won. Note that the coefficient on the first moment in Mùnera's Table II was not normalized to unity as in equation (3), but the ranges above reflect this normalization.

[12]The consistency of linearized moments with gambling and insurance has been pointed out by Mùnera [1985].

TABLE 1. Two examples of Allais paradox

	probability of outcome:					
Gambles	\$0	\$3,000	\$4,000	mean	M_2	M_3
A1	0.2		0.8	3,200	1,600	-1,830
B1		1.0		3,000	0	0
C1	0.8		0.2	800	1,600	1,832
D1	0.75	0.25		750	1,300	1,363
	probability of outcome:					
	\$0	\$40,000	\$100,000			
A2	0.5	0	0.5	50,000	50,000	0
B2		1.0		40,000	0	0
C2	0.99		0.01	1,000	9,950	21,328
D2	0.98	0.02		800	5,600	10,639

comes for each gamble. The table also shows the first three linearized moments of each gamble, allowing the gambles to be evaluated using equation (3). Most experimental subjects will choose gamble B1 over A1, and gamble C1 over D1 (and the same respective choices for the second set of four gambles).[13] These choices suggest inequalities of the form $U(C1) > U(D1)$, which implies $800 + 1600\gamma_2 + 1832\gamma_3 > 750 + 1300\gamma_2 + 1363\gamma_3$.[14] Comparing other choices yields additional constraints, and the region in the $(-\gamma_2, \gamma_3)$ plane that satisfies these constraints is illustrated by the shaded portion in Figure 3.[15] Figure 3 implies that $-\gamma_2 > 0.2$ and $\gamma_3 > .06$. The boundary on γ_2 is very similar to the low end of the range estimated by Mùnera.

[13]The paradox is that for an arbitrary utility function $u(.)$ normalized to $u(0) = 0$, B1 preferred to A1 implies that $u(3000) > 0.8u(4000)$, whereas C1 preferred to D1 implies that $0.2u(4000) > 0.25u(3000) \Rightarrow 0.8u(4000) > u(3000)$.

[14]This exercise assumes a linear cardinal utility function, implying that the expected value of $u(C1)$ is just the expected value of $C1$. Allais [1988] has found that empirical data can be well represented by the utility function $u(W) = log(A_0 + W)$ where A_0 is a very large number. For small to moderate changes in wealth, this function is approximately linear: $u(W)) = log\ A_0 + log(1 + W/A_0) \approx log\ A_0 + W/A_0$.

[15]Inequality constraints were examined for several other versions of the Allais paradox, but these constraints were dominated by the ones shown.

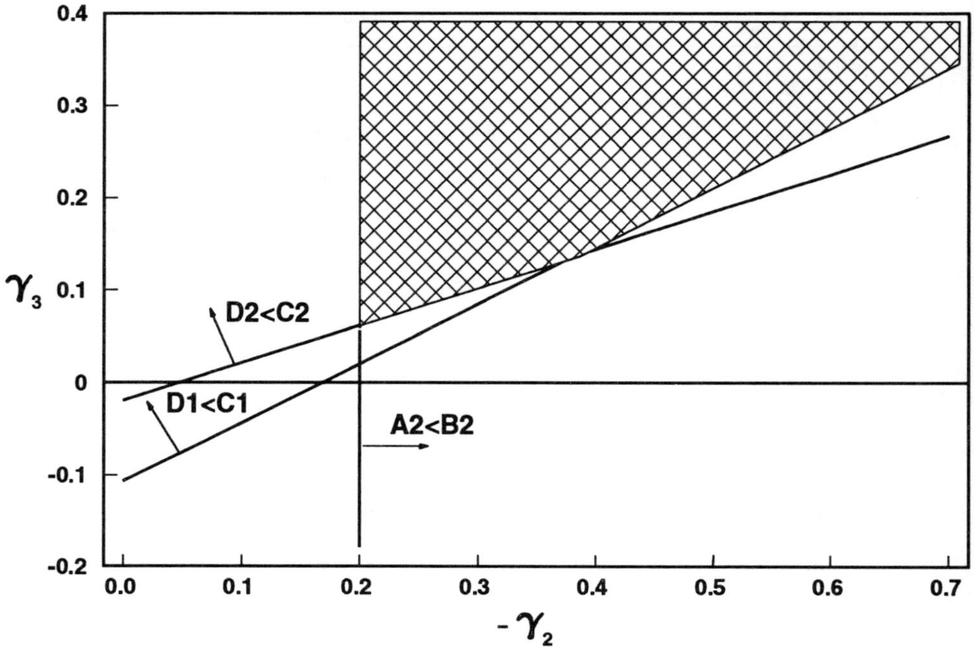

Figure 3. Parameter space implied by Allais Paradox examples (shaded area)

Although the lower bound on γ_3 is also consistent with Mùnera's range, this lower bound is less restrictive than the one estimated by Mùnera.

4. Asset Returns in a Model with Linearized Moments

Having described a static model based on the Allais theory, the next step is to develop a dynamic model and calculate asset prices and returns. This section is divided into three subsections. The first subsection describes the dynamic model and the numerical procedure for solving the model. The second subsection develops expressions for calculating the effect of leverage on the returns to the risky asset. The third subsection calculates asset returns, showing that the model yields risky and risk-free rates consistent with those observed in financial markets.

4.1. A DYNAMIC MODEL

The model in this section is very similar to the Lucas [1978] endowment economy described in Section 2, in which a representative consumer maximizes an objective function subject to a budget constraint. The only change is in the objective function, which becomes a time-separable infinite sum of the static linearized moments model. With all moments defined by time

t expectations, the consumer's time t optimization becomes

$$\begin{aligned} V_t &= \max_{a_t} \sum_{j=0}^{\infty} \beta^j U_t(c_{t+j}), \\ &= \sum_{j=0}^{\infty} \beta^j \left\{ E_t u(c_{t+j}) + \sum_{k=2}^{r} \gamma_k \left\{ E_t \left[u(c_{t+j}) - E_t u(c_{t+j}) \right]^k \right\}^{1/k} \right\}, \quad (4) \end{aligned}$$

$$s.t. \qquad c_t = a_{t-1}(p_t + y_t) - a_t p_t,$$

where c_t, p_t, y_t, and a_t are consumption, asset prices, dividends, and asset holdings at time t, r is the number of linearized moments, and β is the time discount factor.

For notational convenience in writing subsequent equations, M_k and m_k are defined over next period's consumption as follows

$$M_k \equiv \left\{ E_t \left[u(c_{t+1}) - E_t u(c_{t+1}) \right]^k \right\}^{1/k} \equiv \{m_k\}^{1/k}, \qquad k \geq 2.$$

With this notation the partial derivative of M_k with respect to a_t is

$$\frac{\partial M_k}{\partial a_t} = m_k^{1/k-1} E_t m_{k-1} \left[u'(c_{t+1})(p_{t+1} + y_{t+1}) - E_t u'(c_{t+1})(p_{t+1} + y_{t+1}) \right],$$

and the first order condition of equation (4) with respect to a_t leads to the following expression for the price of an asset:

$$\begin{aligned} p_t = {} & \frac{\beta}{u'(c_t)} E_t \left[u'(c_{t+1})(p_{t+1} + y_{t+1}) \right] + \frac{\beta}{u'(c_t)} \sum_{k=2}^{r} \gamma_k m_k^{1/k-1} E_t \{ m_{k-1} \times \\ & \left[u'(c_{t+1})(p_{t+1} + y_{t+1}) - E_t u'(c_{t+1})(p_{t+1} + y_{t+1}) \right] \}. \quad (5) \end{aligned}$$

Only the first two moments are necessary to simulate the desired risk premium, so all higher moments will be ignored in the subsequent derivations and calculations throughout this section. The impact of including the third moment will be discussed briefly in Section 5. Since the model with two linearized moments is very similar to a mean-variance structure that can be equivalent to expected utility, the appendix gives an example that demonstrates an important difference. The appendix shows that a model with two linearized moments is consistent with Allais paradox experiments that violate expected utility.

To further simplify the equations and to accentuate the negative effect of the second moment on prices, γ_2 is replaced by $-\gamma$. The utility functional becomes $U(c_{t+1}) = E_t [u(c_{t+1})] - \gamma M_2$, and equation (5) simplifies to

$$p_t = \frac{\beta}{u'(c_t)}\Big\{E_t\left[u'(c_{t+1})(p_{t+1}+y_{t+1})\right] - \frac{\gamma \text{Cov}_t[u(c_{t+1}), u'(c_{t+1})(p_{t+1}+y_{t+1})]}{M_2}\Big\}, \tag{6}$$

$$= \frac{\beta}{u'(c_t)}\{E_t\left[u'(c_{t+1})(p_{t+1}+y_{t+1})\right] - \gamma\sigma_t[u'(c_{t+1})(p_{t+1}+y_{t+1})]\rho_t[u(c_{t+1}), u'(c_{t+1})(p_{t+1}+y_{t+1})]\},$$

where $\text{Cov}_t(\cdot,\cdot)$, $\rho_t(\cdot,\cdot)$, and $\sigma_t(\cdot)$ are the covariance, correlation, and standard deviations defined by time t expectations.

The first term in the above expression is the same as in the expected utility model, and the second term comes from the second linearized moment. In equation (6), the next period's price plus dividend is always multiplied by the marginal utility of consumption, and this product can be viewed as the "equilibrium valuation" of the asset in the next period. The asset's price relative to the expected utility model is reduced by a term that includes the standard deviation of this valuation and its correlation with the next period's utility. For the cases considered below the correlation term will be positive, so the price of the risky asset will be lower and the returns will be higher than with expected utility.

Next, the cardinal utility function is specified as

$$u(c) = \frac{c^{1-\alpha}}{1-\alpha}, \tag{7}$$

and consumption growth is assumed to follow a Markov process, $c_{t+1} = \lambda_{t+1}c_t$, where λ_{t+1} is unity plus the growth rate from t to $t+1$. For the aggregate asset whose dividends equal total consumption, equation (6) will be linearly homogeneous in dividends. If the price/dividend ratio is defined as $\omega_t = p_t/y_t$, substituting into equation (6) yields equations for the ω's

$$\omega_t = \beta E_t\lambda_{t+1}^{1-\alpha}(\omega_{t+1}+1) - \beta\gamma\sigma_t[\lambda_{t+1}^{1-\alpha}(\omega_{t+1}+1)]\rho_t[\lambda_{t+1}^{1-\alpha}, \lambda_{t+1}^{1-\alpha}(\omega_{t+1}+1)]\,. \tag{8}$$

Assuming q discrete Markov states, $\lambda_t \in \{\lambda_i,\ i = 1,\ldots,q\}$, with transition probabilities $\phi_{ij} = \text{Prob}(\lambda_j|\lambda_i)$, q equations for the q price/dividend ratios can be written as

$$\omega_i = \beta\sum_{j=1}^{q}\phi_{ij}\lambda_j^{1-\alpha}(\omega_j+1) - \beta\gamma\sigma[\lambda_j^{1-\alpha}(\omega_j+1)]\rho[\lambda_j^{1-\alpha}, \lambda_j^{1-\alpha}(\omega_j+1)]. \tag{9}$$

With q linear equations in the unknown ω_i's, numerical solutions can be calculated. The rate of return to the risky equity conditioned on a transition from state i to j can be obtained from the price/dividend ratios

$$r_{ij}^E = \frac{p_j + y_j}{p_i} - 1 = \frac{\lambda_j(\omega_j+1)}{\omega_i} - 1\,. \tag{10}$$

The expected risky rate of return in state i becomes

$$E_i r^E = \sum_{j=1}^{q} \phi_{ij} r_{ij} \ . \tag{11}$$

Equation (6) can also be used to obtain an expression for the risk-free rate. For a one-period bond with dividend equal to unity, substituting $p_{t+1} = 0$, $y_{t+1} = 1$, and continuing as before yields the price of the bond in state i

$$p_i^B = \beta \sum_{j=1}^{q} \phi_{ij} \lambda_j^{-\alpha} - \beta\gamma\sigma(\lambda_j^{1-\alpha})\rho(\lambda_j^{1-\alpha}, \lambda_j^{-\alpha}) \ , \tag{12}$$

and the rate of return is just the inverse of the price minus unity.

The rate of return to a consol that pays a dividend of unity forever can also be derived from equation (6). Substituting $y_{t+1} = 1$ and proceeding as before yields equations in the prices of the consol in each state i

$$p_i^C = \beta \sum_{j=1}^{q} \phi_{ij} \lambda_j^{-\alpha}(p_j^C + 1) - \beta\gamma\sigma[\lambda_j^{1-\alpha}(p_j^C + 1)]\rho[\lambda_j^{1-\alpha}, \lambda_j^{-\alpha}(p_j^C + 1)] \ . \tag{13}$$

The q equations in q unknown prices can be solved numerically, and conditional and expected returns can be calculated using expressions similar to equations (10) and (11).

For the special case of an i.i.d. Markov economy, the above expressions simplify further and the return to the consol becomes equal to that of the one-period bond. The expected equity return and the risk-free rate become respectively

$$Er^E = \frac{E\lambda}{\beta[E\lambda^{1-\alpha} - \gamma\sigma(\lambda^{1-\alpha})]} \tag{14}$$

$$r^B = \frac{1}{\beta\,[E(\lambda^{-\alpha}) - \gamma\sigma(\lambda^{-\alpha})\rho(\lambda^{1-\alpha}, \lambda^{-\alpha})]} \ . \tag{15}$$

4.2. LEVERAGE

The equations derived in Section 4.1 apply to an unleveraged share of economic output. Most publicly held corporations, however, finance some of their assets with debt. Because researchers typically use the S&P 500 stock index as a proxy for a risky asset, many introduce leverage into their Lucas model simulations to capture the effect of debt in the corporate sector. This paper models leverage similar to Benninga and Protopapadakis [1990], except that the debt instrument is a consol that pays one dollar forever rather than a one-period bond. This approach allows the price volatility of the leveraged asset to be more easily defined.

A leveraged corporation finances some of its assets with debt, which commits part of its share of output to interest payments. Because the interest payment is lower than the expected return to equity, the leveraged asset has greater expected return than an unleveraged asset. But with a fixed interest commitment, the net dividend from the leveraged asset has greater volatility.[16]

A share of stock consists of one share of equity, of which θ is financed by selling consols short. With the price of equity and consols as p_t^E and p_t^C, financing θp_t^E requires selling $\theta p_t^E / p_t^C$ consols. The price of this stock at time t will be

$$p_t^S = p_t^E - \left(\frac{\theta p_t^E}{p_t^C}\right) p_t^C = (1-\theta)\, p_t^E. \tag{16}$$

The price of a share of stock at time $t+1$ will be the value of one share of equity minus the value of the debt that was issued at time t

$$p_{t+1}^S = p_{t+1}^E - \left(\frac{\theta p_t^E}{p_t^C}\right) p_{t+1}^C. \tag{17}$$

The net dividend will be the dividend to a share of equity minus the debt's interest payment

$$y_{t+1}^S = y_{t+1}^E - \frac{\theta p_t^E}{p_t^C}. \tag{18}$$

From equations (17) and (18), the return to stock at time $t+1$ will be

$$r_{t+1}^S = \frac{p_{t+1}^S + y_{t+1}^S}{p_t^S} = \frac{r_{t+1}^E - \theta r_{t+1}^C}{1-\theta} \tag{19}$$

where r_{t+1}^E and r_{t+1}^C are the returns to equity and the consol.

4.3. EQUITY PREMIUM CALCULATIONS

This section calculates asset returns using the same exogenous per capita consumption series used by several others in equity premium simulations. Annual growth in per capita consumption has a mean, standard deviation, and autocorrelation of 1.8%, 3.6%, and –0.14 respectively. With a two-state Markov process, the exogenous parameters for the equations in Section 4.1 are $\lambda_1 = 1.054$, $\lambda_2 = 0.982$, $\phi_{11} = \phi_{22} = 0.43$, and $\phi_{12} = \phi_{21} = 0.57$.[17]

[16] Note that the approach used in this paper does not allow for the possibility of bankruptcy, which was considered by Kandel and Stambaugh [1991] and Cecchetti, Lam, and Mark [1993]. Since there is a lower limit to the dividend of the unleveraged asset in a discrete state economy, bankruptcy can be avoided in the model by keeping the leverage from getting too large in any period.

[17] Mehra and Prescott [1985] found that their result was not sensitive to the number of states.

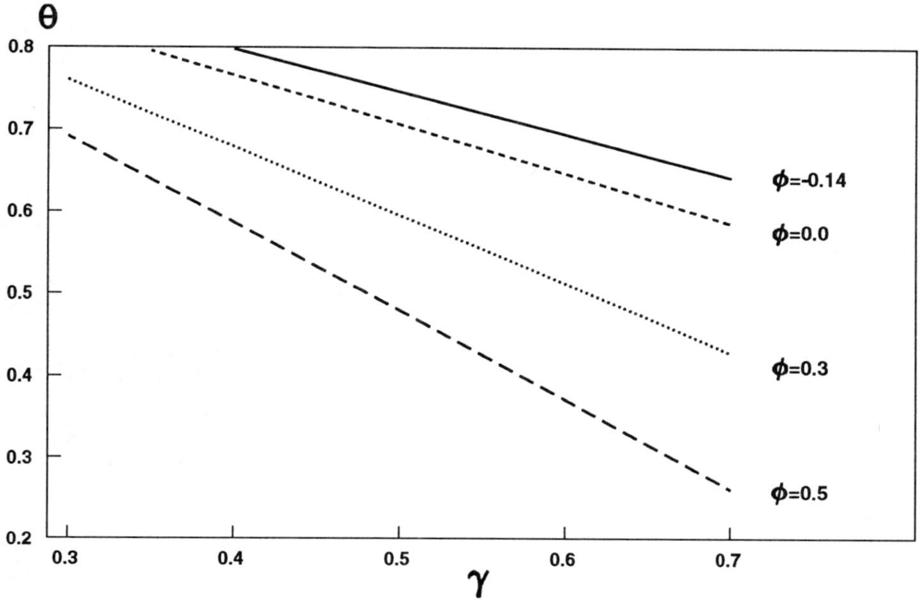

Figure 4. Risk (γ) and leverage (θ) parameters that yield 6.18% risky return when risk-free rate = 0.8% (ϕ is the autocorrelation in consumption growth)

The cardinal utility function has little curvature according to Allais's [1988] estimate, so the first case calculated is $\alpha = 0$, i.e. $u(c) = c$. Under this assumption asset demand will not be uniquely defined in the equation (4) optimization because consumption and assets are perfect substitutes. Nevertheless, the expressions for prices and returns are still valid at an equilibrium where the representative agent holds the aggregate asset. Results have also been calculated for α varying from zero to one, and the results do not change significantly over this range of utility curvature.

For linear cardinal utility, the time discount parameter β is uniquely identified by the desired risk-free rate. The historical 0.8% interest rate implies that $\beta = 0.992$, which is within the range of values estimated by other researchers (Hansen and Singleton, 1983). The two remaining parameters in the model are the risk (γ) and leverage (θ) parameters. These parameters are not uniquely identified by the mean asset returns. A range of (γ, θ) pairs are consistent with the 6.98% risky return targeted by Mehra and Prescott.

Figure 4 graphs (γ, θ) pairs that yield the desired 6.98% risky return. Since asset returns can be sensitive to the time series properties of the consumption process, Figure 4 shows solutions for several values for the autocorrelation in consumption growth (ϕ). The values –0.14 and 0.0 are

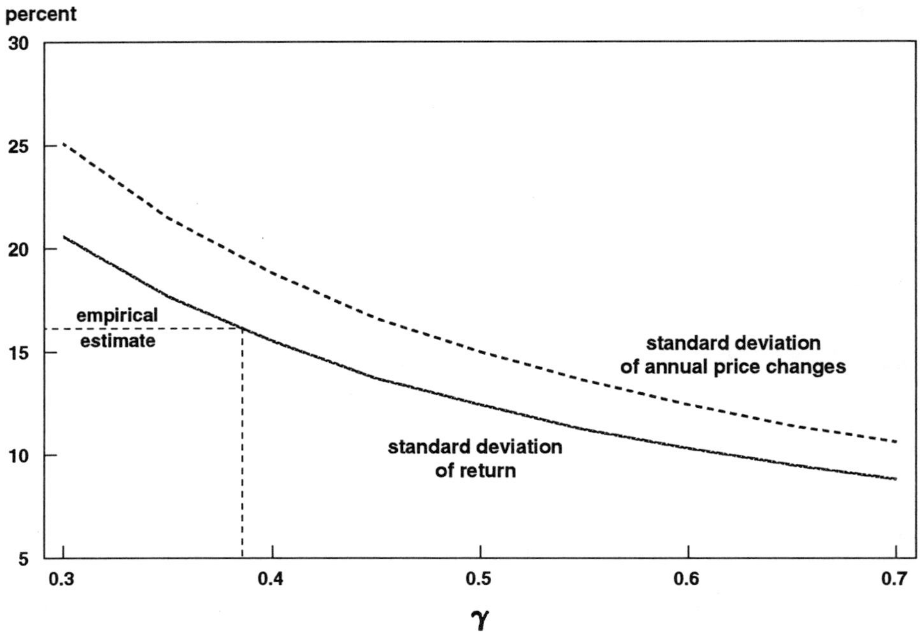

Figure 5. Volatility of risky return and price changes for $\phi = 0.3$

chosen for comparison with other work while the positive values 0.3 and 0.5 correspond to the respective estimates for the 1948-1990 and 1930-1990 intervals.[18] Although the equity premium tends to increase as ϕ becomes more positive, the desired asset returns can be simulated for all values of ϕ shown. Note that the risk parameter γ is in the range estimated by Mùnera [1988], and the leverage parameter is in a range that others have considered reasonable.

Although mean asset returns do not uniquely identify the risk parameter, additional constraints on the model can be imposed by considering the second moment of returns. Figure 5 plots the standard deviation of both returns and percentage price changes for the risky asset as a function of the risk parameter γ, where the other parameters are chosen to generate the desired mean returns. A risk parameter of $\gamma \approx 0.39$ yields the 16.5% standard deviation of returns calculated by Mehra and Prescott from historical data.

Considering the second moment of the risk-free rate reveals a potential

[18]The robustness of the results to changing consumption autocorrelation is a positive feature of the model. Golob [1992] has shown that some "solutions" to the puzzle are very sensitive to the specification of the consumption process.

weakness of the model. For linear cardinal utility the risk-free rate is constant, compared to a standard deviation of 5.7% in the Mehra-Prescott time interval. By increasing the curvature in the utility function, the standard deviation of the risk-free rate increases to about one percent in the model. Although this is representative of the U.S. post-World War II data, it is substantially less than the 5.7% standard deviation in the Mehra-Prescott 1889-1978 period.

The absence of money from the Lucas endowment economy is one possible explanation for low volatility of the risk-free rate in the model. Unexpected inflation implies that the risk-free rate calculated ex post differs from that expected ex ante. This would induce volatility in the risk-free rate that would not be captured by a model without money. Also, monetary policy could affect the risk-free rate over the short-term through a liquidity effect (Christiano and Eichenbaum, 1992). Thus, to the extent that inflation and monetary policy increase volatility in the risk-free rate, a model without money would be expected to have less volatility.

5. Discussion and Conclusion

The calculations in Section 4.3 consider only the first two moments of the exogenous consumption process, but the model can easily be extended to include the third moment. Agents favor positive skewness, so negative skewness in per capita consumption will reduce the price of the risky asset, which leads to an increase in the risky return. Nevertheless, a model with three linearized moments can replicate historical asset returns, provided that either the leverage or the other risk parameter is reduced relative to the two moment solution. All parameters are still within "reasonable" bounds in the three-moment simulation (not shown).

The calculations in Section 4.3 also assume that cardinal utility is linear, but the expressions in Section 4.1 are general enough to consider the nonlinear case. While not shown, these calculations indicate that neither the risky nor risk-free asset return changes substantially when the curvature parameter α is varied between zero and one. Thus, results are only shown for the simple form of the model.

To summarize, this paper extends a static linearized moments preference structure into a dynamic Lucas model. Linearized moments preferences are consistent with many static choice experiments that are inconsistent with expected utility, including the Allais paradox. The model is also consistent with the simultaneous purchase of lotteries and insurance, whereas the Friedman-Savage explanation based on expected utility fails to explain this result over a wide distribution of wealth. Finally, the dynamic version of the model replicates the risky and risk-free returns in historical data.

Some authors have suggested that solutions to the equity premium puzzle will require relaxing the complete markets hypothesis (Weil, 1989; Kocherlakota, 1995), but this paper suggests that agent preferences are the key to the puzzle.

The paper also shows how the Allais criticism of expected utility can explain the equity premium puzzle. When the risk aversion coefficient in expected utility is increased, the marginal utility of consumption becomes so volatile that the risk-free asset becomes implicitly risky. The essence of the Allais solution is to separate risk from marginal utility.

Compared to many equity premium simulations, an important feature of the linearized moments approach is the consistency of its risk parameters with estimates from static choice experiments. Microeconomic foundations are an important motivation for dynamic macroeconomic models, so parameter estimates should be consistent with micro data.

A. Appendix: Example where "two linearized moments" model is consistent with choices that violate expected utility

The following table reproduces four gambles from Kahneman and Tversky (1979), and their means and standard deviations.

Gamble			μ	σ
A:	Prob($4000)=0.8	Prob($0)=0.2	3200	1600
B:	Prob($3000)=1.0		3000	0
C:	Prob($4000)=0.2	Prob($0)=0.8	800	1600
D:	Prob($3000)=0.25	Prob($0)=0.75	750	1300

Typical preferences (for 60-70% of experimental subjects) are $A \prec B$ and $C \succ D$. Inconsistency with expected utility can be seen by noting that $A \prec B \Rightarrow 0.8U(4000) < U(3000)$, but $C \succ D \Rightarrow 0.2U(4000) > 0.25U(3000) \Rightarrow 0.8U(4000) > U(3000)$.

While these choices violate expected utility, they can be consistent with preferences of the form $U = \mu - \gamma\sigma$, where μ and σ are the respective mean and standard deviation of the gamble, and γ is a constant. For this model:
$A \prec B \Rightarrow 3200 - 1600\gamma < 3000 \Rightarrow \gamma > 1/8$, and
$C \succ D \Rightarrow 800 - 1600\gamma > 750 - 1300\gamma \Rightarrow \gamma < 1/6$.

Thus, for $1/8 < \gamma < 1/6$, $U = \mu - \gamma\sigma$ is consistent with the Kahneman-Tversky gambles that violate expected utility.

References

Abel, A. B. (1991), "The Equity Premium Puzzle," *Business Review*, Federal Reserve Bank of Philadelphia, 1991:5, 3-13.

Allais, M. (1952), "Fondements d'une Théorie Positive des Choix comportant un risque et Critique des Postulats et Axiomes de l'Ecole Américaine," in Econométrie, Colloques Internationaux du Centre National de la Recherche Scientificque, Vol. XL, Paris, 257-332, Translated in Allais and Hagan (1979) as: "The Foundations of a Positive Theory of Choice involving Risk and a Criticism of the American School, 27-145.

Allais, M. (1988), "The General Theory of Random Choices in Relation to the Invariant Cardinal Utility Function and the Specific Probability Function. The (U,θ) Model: A General Overview," in *Risk, Decision, and Rationality*, B. R. Munier (ed), Dordrecht: Reidel.

Allais, M. and Hagen, O. (eds.) (1979), *Expected Utility Hypotheses and the Allais Paradox*, Dordrecht: Reidel.

Baumol, W. J. (1958), "The Cardinal Utility which is Ordinal," *Economic Journal*, 68, 665-72.

Benninga, S. and Protopapadakis, A. (1990), "Leverage, Time Preference, and the Equity Premium Puzzle," *Journal of Monetary Economics*, 25, 49-58.

Cecchetti, S. G., Lam, P., and Mark, M. C. (1993), "The Equity Premium and The Risk Free Rate: Matching the Moments," *Journal of Monetary Economics*, 31, 21-45.

Christiano, L. J. and M. Eichenbaum (1992), "Liquidity Effects, Monetary Policy, and the Business Cycle," *Economic Perspectives*, Federal Reserve Bank of Chicago, November-December, 2-14.

Epstein, L. G. and Zin, S. E. (1989), "Substitution, Risk Aversion, and the Temporal Behavior of Consumption and Asset Returns: A Theoretical Framework," *Econometrica*, 57:4, 937-969.

Fishburn, P. C. (1988), *Nonlinear Preference and Utility Theory*, Baltimore: Johns Hopkins Press.

Friedman, M. and Savage, L. J. (1948), "The Utility Analysis of Choices Involving Risk," *Journal of Political Economy*, 56:4, 279-304.

Golob, J. E. (1992), "A Regime Shift in Per Capita Consumption, with Implications for Asset Prices and Returns," Working Paper 92-08, Federal Reserve Bank of Kansas City.

Grossman, S. and Shiller, R. (1981), "The Determinants of the Variability of Stock Prices," *American Economic Review*, 71, 222-227.

Hagen, O. (1988), "Expected Utility Theory and Ordinalism. A Political Marriage," in *Risk, Decision, and Rationality*, B. R. Munier (ed), Dordrecht: Reidel.

Hansen, P. L. and Singleton, K. J. (1983), "Stochastic Consumption, Risk Aversion, and the Temporal Behavior of Asset Returns," *Journal of Political Economy*, 91, 249-265.

Hirshleifer, J. (1970), *Investment, Interest, and Capital*, Englewood Cliffs: Prentice-Hall.

Kahneman, D. and Tversky, A. (1979), "Prospect Theory: An Analysis of Decision Under Risk," *Econometrica*, 47, 263-292.

Kandel, S. and Stambaugh, R. F. (1991), "Asset Returns and Intertemporal Preferences," *Journal of Monetary Economics*, 27, 39-71.

Kocherlakota, N. R. (1995), "The Equity Premium: It's Still a Puzzle," Institute for Empirical Macroeconomics, Federal Reserve Bank of Minneapolis, Discussion Paper 102.

Kreps, D. M. and Porteus, E. L. (1978), "Temporal Resolution of Uncertainty and Dynamic Choice Theory," *Econometrica*, 36, 185-200.

Lucas, R. E. (1978), "Asset Prices in an Exchange Economy," *Econometrica*, 46, 1429-1446.

Markowitz, H. (1952), "The Utility of Wealth," *Journal of Political Economy*, 60, 151-58.

Marshall, A. (1898), *Principles of Economics*. London: Macmillan.

Mehra, R. and Prescott, E. C. (1985), "The Equity Premium: A Puzzle," *Journal of Monetary Economics*, 15, 145-161.

Munier, B. R. (ed.), (1988), *Risk, Decision and Rationality*, Dordrecht: Reidel.

Munier, B. R. (1989), "New Models of Decision under Uncertainty: An Interpretative Essay," *European Journal of Operational Research*, 38, 307-317.

Mùnera, H. H. (1985), "The Generalized Means Model (GMM) for Non- Deterministic Decision Making; Its Normative and Descriptive Power, Including Sketch of the Representation Theorem," *Theory and Decision*, 18, 173-202.

Mùnera, H. H. (1988), "A Large Scale Empirical Test for the Linearized Moments Model (LMM): Compatibility Between Theory and Observation," in *Risk, Decision, and Rationality*, Dordrecht: Reidel.

M'unera, H. A. and de Neufville, R. (1983), "A Decision Analysis Model when the Substitution Principle is Not Applicable," in B. P. Stigum and F. Wenstøp (eds.), *Foundations of Utility and Risk Theory with Applications*, 247-262.

Schoemaker, P. J. H. (1982), "The Expected Utility Model: Its Variants, Purposes, Evidence and Limitations," *Journal of Economic Literature*, 20, 529-563.

Stigum, B. P. and Wenstøp, F. (eds.) (1983), *Foundations of Utility and Risk Theory with Applications*, Dordrecht: Reidel.

Weber, M. and C. Camerer. (1987), "Recent Developments in Modelling Preferences under Risk," *OR Spektrum*, 9, 129-151.

Weil, P. (1989), "The Equity Premium Puzzle and the Riskfree Rate Puzzle," *Journal of Monetary Economics*, 24, 401-421.

Weil, P. (1990), "Nonexpected Utility in Macroeconomics," *Quarterly Journal of Economics*, 105, 29-42.

GENERALIZED EXPECTED UTILITY AND THE DEMAND FOR INSURANCE
The Limits of Machina

THOMAS RUSSELL
Dept. of Economics
Santa Clara University
Santa Clara, CA 95053

1. Introduction

Under certain circumstances, every smooth functional admits a 'Taylor series' expansion in polynomials which, as we raise the order of the polynomial, provides successively more accurate approximations to the value of the functional in a neighborhood of some point in function space.

This fact provides the mathematical foundation for Machina's generalized expected utility approach to decision making under uncertainty, Machina [1982]. The purpose of this paper is to revisit this approach from two different perspectives.

In the first place we will argue that for those decision problems to which the Machina approach can be applied, it has been applied too timidly. Since the accuracy of the Taylor series approximation is improved by adding successively higher polynomials in probability, there seems no reason (other than deference to the Bernoulli doctrine) to limit oneself to the first (linear) term in the Taylor series expansion.

In the second place we will argue that there are a number of important economic problems to which the approach has been applied too boldly. For a functional to admit a Taylor expansion which is polynomial in probabilities, the cumulative probability distributions over which it is defined must all have the same support. This requirement is frequently not met (for example, it is not met in many models of the individual demand for insurance) and in these cases no linear expansion in probabilities is available.

To place these points in historical perspective, we begin with a brief history of the Taylor series approach to functional approximation.

2. Polynomial Expansion of Functionals: a Brief History

As soon as one thinks of a function as a point in a suitably structured space, it becomes very natural to ask whether or not the famous Weierstrass Theorem - that every function on a compact set can be approximated arbitrarily closely by a polynomial - may not have a higher analysis counterpart that every functional defined on a suitable space of

R. Nau et al. (eds.), Economic and Environmental Risk and Uncertainty, 109–114.

functions has a polynomial approximation in some sense.

The possibility of such a Weierstrass result seems first to have been noted by Frechet. A proof of the proposition was given by Gateaux, but the result only seems to have become widely known as a consequence of the lectures of Levy [1922]. At Chapter 6, Levy points out that for functions defined on the interval [0, 1], the pth term in the polynomial expansion of any smooth functional is of the form

$$\int_0^1 \int_0^1 \ldots \int_0^1 \phi(t_1, \ldots t_p)\, x(t_1) \ldots x(t_p)\, dt_1 \ldots dt_p . \tag{1}$$

In Machina's context the first order term in (1) is just local expected utility.

Levy's lectures proved to be extremely timely, and were used extensively in the then emerging field of quantum mechanics. (The mathematical relationship between the fields of quantum mechanics and decision making under uncertainty is very deep, it being no accident that both fields were axiomatised by von Neumann.) These same ideas might have also been expected to play a role in the new field of mathematical statistics, where the need to expand a functional on a space of probability distribution functions was first noted by R. von Mises [1947], but this literature shows no awareness of either Levy's book or the ideas of quantum mechanics.

The ideas of von Mises, however, were eventually related to the standard functional analysis literature by Reeds [1976], see also Fernholz [1983], and Reeds' discussion provides a clear mathematical foundation for a general theory of choice under uncertainty.

Recall that any transformation which maps cumulative probability distributions to the real line is called a statistical functional. For any such functional, say T, and for any two probability distribution functions, say F and G, we may construct the function A of the real variable t defined by

$$A(t) = T((1-t)F + tG) = T(F + t(G - F)) \tag{2}$$

The values of $\dfrac{d^k A(t)}{dt^k}$ evaluated at $t = 0$, $k = 1, 2, \ldots 3$, then provide the coefficients of what Reeds calls the von Mises functional of degree k at F if, for each G, the derivatives $\dfrac{d^i A(t)}{dt^i}\Big|_{t=0}$ of $A(t) = T(F + t(G - F))$ exist for $i = 1, \ldots k$ and if there are functions $a_1, a_2, \ldots a_k$ such that for $i = 1, \ldots k$

$$\frac{d^i A(t)}{dt^i}\Big|_{t=0} = \int \ldots \int a_i(x_1, \ldots x_i)\, d(G-F)(x_1) \ldots d(G-F)(x_i) \tag{3}$$

The functions $a_1, a_2 \ldots$ are called the kernel functions of T at F. The expansion of a functional around F is then given by

$$T(G) = T(F) + \int a_1(x)\,d(G-F)(x) +$$

$$\frac{1}{2}\int\int a_2(x,y)\,d(G-f)(x)\,d(G-f)(y) +$$

$$\ldots\frac{1}{k!}\int\ldots\int a_k(x_1,\ldots x_k)\,d(G-F)(x_1)\ldots d(G-F)(x_k) \tag{4}$$

We turn now to examine how this expansion may be used in the theory of decision making under uncertainty.

3. Generalizing General Expected Utility

These results of Reeds in the context of Machina's approach suggest a strategy for modeling choice under uncertainty. Assuming that the conditions required to be able to expand an arbitrary smooth choice functional are met, take as many terms in the von Mises expansion as are necessary to fit whatever data needs to be explained. Practically speaking it seems unlikely that one would wish to go far beyond the quadratic approximation, so we might consider modeling choice which is governed by a general functional V by replacing V by a functional of the form

$$T = \int a_1(x)\,d(G-F) + \frac{1}{2}\int\int a_2(x,y)\,d(G-F)(x)\,d(G-F)(y) \tag{5}$$

This will generally provide a reasonable approximation to arbitrary smooth functionals in a neighborhood of F. Here the kernel functions a_1, a_2 are the first and second derivatives of the functional V. The derivative is defined in the Hadamard or compact sense as explained by Reeds, not in the Frechet sense as used by Machina. In this context, the Frechet derivative often will not exist.

The theory of quadratic functionals, i.e. the definition of risk aversion, the analysis of comparative statics and so on, needs to be worked out. Perhaps a quadratic functional will not fit the data, but in this case one would simply add a term to the expansion knowing that this can only improve the fit. Thus this approach systematically expands on general expected utility in a way which can only give better approximation.

4. The Limits of the Frechet-Gateaux-von Mises-Machina Approach

It is straightforward to extend the thrust of Machina's paper by including higher order terms in the polynomial expansion of the valuation functional. As with any theorem, however, the 'Taylor series for functionals' expansion theorem requires assumptions. (See Allen [1987] for a discussion of sufficient conditions in the context of Machina.) In particular, to be able to expand the functional in polynomials, it is necessary that the distributions over which the functional is defined all have the same support.

To see why this condition is necessary, consider the following simple choice problem. Suppose there are two states of nature, and suppose, therefore, that the individual's choice problem has three parameters, P_1, the probability that state 1 will occur, (P_2, the probability of state 2 being $(1 - P_1)$), X_1, the payoff if state 1 occurs, and X_2, the payoff if state 2 occurs.

In general we may assume the individual has a valuation function given by $V(P_1, X_1, X_2)$. There are two cases to consider.

When the individuals choices affect probabilities but not outcomes, we have $P_1 = P_1(\alpha)$ where α is a choice parameter. In this case elementary calculus of several variables shows that the expansion of $V(\alpha)$ in a Taylor series gives a first term which is linear in probability, a second term which is quadratic in probability and so on.

Secondly, when the individual agent's choices affect outcomes but not probabilities, we have $X_1 = X_1(\beta)$, $i = 1, 2$, where β is the choice parameter. Now $V(\beta) = V(P, X_1(\beta), X_2(\beta))$ can be expanded in a Taylor series whose first term is linear in X_i, but, in general this cannot be expected to be linear in P. There is no first order expected utility approximation in this case.

It seems, then, that the Machina approach can only be applied when the choice problem involves changes in probabilities on a given support. Now there are many economic problems in which choice behavior does change the probability distributions over a fixed support so that the approach is applicable. An obvious example is portfolio choice theory in finance. In that case Taylor series expansions lead to a natural geometric structure in which the terms of the Taylor series are elements of the jet space of a suitable manifold, see Russell [1990].

There are also many economic problems, however, where choice does not affect probability but only outcomes, and here the approach cannot be used. An obvious example is the classic problem of individual demand for insurance. This problem can be formulated as follows. (We follow Karni [1992]).

Let X be a random variable representing possible losses of wealth in case of an accident, and let F be its cumulative distribution function. The support of F is [0, w] where w is initial wealth. Denote by $I : R \rightarrow R$ the indemnity function giving the

insurance payoff in the event of an accident. Let P be the insurance premium. Assume that $P = \chi \int I(x)\, dF(x)$
where χ is a loading factor and x is the outcome of X. Terminal wealth of the insured is now given by the random variable Y, where

$$Y(X) = w - P - X + I(x)$$

In this case the choice of an indemnity function I (x) changes the payoffs to the individual in the event of an accident, but it does not change the probability of the accident. This is given by nature. Although it may appear that all distributions are defined on the common support [0, w], in fact one of the primary purposes of the purchase of insurance is to change this support. For example the uninsured agent may well face positive probability mass at zero wealth, but after the purchase of insurance, the probability distribution may place no weight on outcomes below some wealth level L. This action then changes the support of the distribution. In this case, given some valuation functional V, we can obviously calculate a Taylor series expansion, but the first order term will not in general be linear in probability. Any non-linearities in probabilities remain to first order. Thus if we wish to develop a non-Bernoulli theory to explain demand for insurance, the use of the general expected utility **local expected utility** function must be justified on some grounds other than the usual argument that for small changes in probabilities all functionals are locally Bernouillian.

5. Conclusion

The Frechet-Gateaux approach to functional approximation as developed in statistics by von Mises and in economics by Machina is an extremely powerful tool providing functional forms which locally approximate as closely as one wishes any functional representing smooth transitive preferences. Low order polynomials, say quadratics, are a natural candidate for fitting to data, and a quadratic expansion plays somewhat the same role in the economics of choice under uncertainty that Diewert's [1974] flexible functional forms play in the theory of choice under certainty.

On the other hand this approach does have its limits. Many choice problems in economics do not involve choice of a distribution function from a set of functions with a common support. Whenever the individual is given choices of payoffs which change the support of the distributions, the first order approximations are not linear in probabilities, so local expected utility theory will not apply.

6. Bibliography

Allen, B. (1987) "Smooth Preferences and the Approximate Expected Utility Hypothesis," *Journal of Economic Theory,* **41**, 340-355.

Diewert, I. (1974) "Applications of Duality Theory," in M Intriligator and D Kendrick (eds.) *Frontiers of Quantitative Economics,* Amsterdam: North Holland pp. 106-171.

Fernholz, L.T. (1983) *Von Mises Calculus for Statistical Functionals.* New York: Springer Verlag.

Karni, E. (1992) "Optimal Insurance: A Nonexpected Utility Analysis," in G Dionne (ed.) *Contributions to Insurance Economics,* Boston: Kluwer, pp. 217-237.

Levy, P. (1922) *Lecons d'Analyse Fonctionelle.* Paris: Gauthier-Villars.

Machina, M. (1982) "Expected Utility Without the Independence Axiom," *Econometrica,* **50,** 277-323.

von Mises, R. (1947) "On the Asymptotic Distribution of Differentiable Statistical Functions," *Annals of Math Stat.,* **21,** 309-348.

Reeds, J. A. (1976) "On the Definition of von Mises Functionals," *Ph.D. Harvard University.*

Russell, T. (1990) "Choice as Geometry," in R Sato and R Ramachandran (eds) *Conservation Laws and Symmetry,* Boston Kluwer pp. 175-224.

THE DERIVATION OF GENERALIZED EXPECTED UTILITY EXPANSIONS

Reply to Russell

MARK J. MACHINA[1]
Department of Economics
University of California, San Diego
La Jolla, California 92093-0508, U.S.A.

In Machina (1982), I presented an analytical approach, termed "generalized expected utility analysis," for the study of smooth non-expected utility preference functionals $V(\cdot)$ defined over cumulative distribution functions $F(\cdot)$ on an outcome interval $[0,M]$. Specifically, I showed that:

> If such a functional is differentiable in $F(\cdot)$ in a formally specified sense, it will possess a first order linear approximation – a "local expected utility" differential expansion involving a "local utility function" $U(\cdot;F)$ – at each distribution $F(\cdot)$.
>
> In such a case, the method of path integration can be applied to obtain exact results on how properties of the local utility functions are linked to global properties of the preference functional – or in other words, to properties of preferences over large changes in distributions, and hence to global attitudes toward risk.

In his critique, Professor Russell (1996) makes the following assertions:

#1 "for those decision problems to which the Machina approach can be applied, it has been applied too timidly. Since the accuracy of the Taylor series approximation is improved by adding successively higher polynomials in probability, there seems no reason ... to limit oneself to the first (linear) term in the Taylor series expansion."

#2 On the other hand, there are cases where "the approach has been applied too boldly. For a functional to admit a Taylor expansion which is polynomial in probabilities, the cumulative probability distributions over which it is defined must all have the same support. This requirement is frequently not met (for example, it is not met in many models of the individual demand for insurance) and in these cases no linear expansion in probabilities is available." Russell argues this by considering two insurance problems, and claiming that my approach cannot be applied to either of them.

---

[1] I am grateful to Mark Durst and Joel Sobel for helpful discussions on this material, and to the National Science Foundation (Grant SES 92-09012) for financial support.

R. Nau et al. (eds.), Economic and Environmental Risk and Uncertainty, 115–122.

The purpose of this paper is to argue that:

Assertion #1 seems to reflect a misunderstanding of my approach: It confuses Taylor Series *approximations* of functions or functionals with the *exact results* implied by the use of path integration and the Fundamental Theorem of Calculus.

Assertion #2 seems to reflect another a misunderstanding: It confuses expansions with respect to small changes in a probability distribution's *outcomes* with expansions with respect to small changes in its *cumulative distribution function* $F(\cdot)$, which form the basis of Machina (1982). By expressing Russell's two insurance problems in terms of preferences over cumulative distribution functions, I illustrate how to apply my approach to obtain the first order local expected utility expansion for each of them.

Section 1 presents a brief but self-contained description of the approach of Machina (1982), and Sections 2 and 3 respond to each assertion.

1. Differentiable Preferences and Local Utility Functions

The approach of Machina (1982) begins with the choice set $D[0,M]$ of all cumulative distribution functions $F(\cdot)$ over a real outcome interval $[0,M]$. This set is endowed with the weak convergence topology (e.g., Billingsley (1968)), which I show (p.314, Lemma 1) to be induced by the norm

$$\|F^*-F\| \equiv \int_0^M |F^*(x)-F(x)|\cdot dx \tag{1}$$

I assume that the individual's preferences over this choice set can be represented by a real-valued preference functional $V(\cdot)$, which is continuous in the weak convergence topology, and differentiable in the sense that at each $F(\cdot) \in D[0,M]$, there exists a continuous linear functional $\psi(\cdot;F)$ defined over the difference space $\Delta D[0,M] = \{\lambda\cdot(F^*-F)\,|\,F^*, F \in D[0,M]; \lambda \in R^1\}$ such that

$$V(F^*)-V(F) = \psi(F^*-F;F) + o(\|F^*-F\|) \tag{2}$$

where $o(\cdot)$ denotes a function that is 0 at 0, of higher order than its argument, and whose convergence to 0 is uniform in $\|F^*-F\|$. By means of the Hahn-Banach Theorem, the Riesz Representation Theorem, and a formally proven lemma (p.314, Lemma 2), I show that this implies

$$V(F^*)-V(F) = \int_0^M U(x;F)\cdot[dF^*(x)-dF(x)] + o(\|F^*-F\|) \tag{3}$$

for some absolutely continuous function $U(\cdot;F)$ over $[0,M]$. Since the first order term (the integral term) in (3) can be interpreted as the change in the expectation of the "local utility function" $U(\cdot;F)$ at $F(\cdot)$, with respect to the change from $F(\cdot)$ to $F^*(\cdot)$, such smooth preference functionals can be interpreted as "locally expected utility."

The fundamental approach of Machina (1982), used in all six of its theorems, is the application of local utility functions to the exact, global analysis of the preference functional $V(\cdot)$ by means of integrals over paths of the form $\{F(\cdot;\alpha)\,|\,\alpha \in [0,1]\} \subseteq D[0,M]$. Specifically, if such a path is "smooth enough" so that $\|F(\cdot;\alpha)-F(\cdot;\alpha_0)\|$ is differentiable in α at $\alpha=\alpha_0$, then (3) implies

$$\begin{aligned}\frac{dV(F(\cdot;\alpha))}{d\alpha}\bigg|_{\alpha=\alpha_0} &= \frac{d}{d\alpha}\left(\int_0^M U(x;F(\cdot;\alpha_0))\cdot dF(x;\alpha) + o\big(\|F(\cdot;\alpha)-F(\cdot;\alpha_0)\|\big)\right)\bigg|_{\alpha=\alpha_0}\\ &= \frac{d}{d\alpha}\left(\int_0^M U(x;F(\cdot;\alpha_0))\cdot dF(x;\alpha)\right)\bigg|_{\alpha=\alpha_0}\end{aligned} \tag{4}$$

so that applying the Fundamental Theorem of Integral Calculus to the path $\{F(\cdot;\alpha)\,|\,\alpha\in[0,1]\}$ yields

$$V(F(\cdot;1)) - V(F(\cdot;0)) = \int_0^1\left(\frac{d}{d\alpha}\left(\int_0^M U(x;F(\cdot;\alpha_0))\cdot dF(x;\alpha)\right)\bigg|_{\alpha=\alpha_0}\right)\cdot d\alpha_0 \tag{5}$$

Equation (5) illustrates how the evaluation of the global change from distribution $F(\cdot;0)$ to $F(\cdot;1)$ can be exactly expressed in terms of the characteristics of the local utility functions $\{U(\cdot;F(\cdot;\alpha))\,|\,\alpha\in[0,1]\}$ at the distributions $F(\cdot;\alpha)$ along the path from $F(\cdot;0)$ to $F(\cdot;1)$.

Some of the applications of generalized expected utility analysis, as developed in Machina (1982), include the following characterizations of global risk attitudes for general "smooth" preference functionals, which closely parallel their corresponding expected utility results:

> $V(\cdot)$ exhibits *global first order stochastic dominance preference* if and only if $U(x;F)$ is nondecreasing in x, at each distribution $F(\cdot) \in D[0,M]$.
>
> $V(\cdot)$ exhibits *global risk aversion* if and only if $U(x;F)$ is concave in x, at each distribution $F(\cdot) \in D[0,M]$.
>
> A preference functional $V^*(\cdot)$ with local utility function $U^*(x;F)$ is *globally at least as risk averse* as $V(\cdot)$ (in a set of formally specified senses) if and only if $U^*(x;F)$ is at least as concave in x as $U(x;F)$, at each distribution $F(\cdot) \in D[0,M]$.

Some examples of preference functionals which are everywhere differentiable[2] over $D[0,M]$, and their corresponding local utility functions, are given in the following table. As such, these preference functionals are subject to analysis by means of equations (4) and (5), and exhibit the above characterizations of global risk attitudes in terms of the properties of their local utility functions:

---

[2] Given the obvious regularity conditions on the component functions $\upsilon(\cdot)$, $\tau(\cdot)$, $\psi(\cdot,\ldots,\cdot)$, etc.

Preference Functional $V(F)$	Local Utility Function $U(x;F)$	Reference
$\dfrac{\int_0^M \upsilon(x)\cdot dF(x)}{\int_0^M \tau(x)\cdot dF(x)}$	$\dfrac{\int_0^M [\upsilon(x)\cdot\tau(\omega)-\tau(x)\cdot\upsilon(\omega)]\cdot dF(\omega)}{\left[\int_0^M \tau(\omega)\cdot dF(\omega)\right]^2}$	Chew (1983)
$\int_0^M\int_0^M \nu(x,y)\cdot dF(x)\cdot dF(y)$	$\int_0^M [\nu(x,\omega)+\nu(\omega,x)]\cdot dF(\omega)$	Chew, Epstein and Segal (1991)
$\psi(\kappa_1(F),\ldots,\kappa_n(F))$ with $\kappa_i(F) = \int_0^M x^i\cdot dF(x)$	$\sum_{i=1}^{n}\dfrac{\partial\psi(\kappa_1(F),\ldots,\kappa_n(F))}{\partial\kappa_n}\cdot x^i$	Bardsley (1993)

See Machina (1982) for further details and discussion of generalized expected utility analysis, and for formal proofs of the above characterizations. This analytical approach has been extended to larger classes of preference functionals and distributions by Bardsley (1992), Chew, Karni and Safra (1987), Karni (1987, 1989) and Wang (1993); it has been formally axiomatized (for finite outcome sets rather than $[0,M]$) by Allen (1987); and it has been applied to analyses of choice under uncertainty by Chew, Epstein and Zilcha (1988), Chew and Nishimura (1992), Green and Jullien (1988), Machina (1984), Machina and Neilson (1987), Röell (1986) and others.

2. Taylor Series Approximations vs. the Fundamental Theorem of Calculus

Russell's first assertion is later followed by the suggestion that "It is straightforward to extend the thrust of Machina's paper by including higher order terms in the polynomial expansion of the valuation functional."

As mentioned, both the assertion and this suggestion seem to reflect a misunderstanding of the main point of my approach. The key idea is *not* to use equations (2) and (3) merely to obtain an approximation to the preference functional $V(\cdot)$ – if that had been my point, I certainly agree that quadratic and higher order terms could (and should) be added. Rather, the point of my approach is to use equations (4) and (5) – *path integrals of first order approximations* – to obtain an *exact* representation of global changes in $V(\cdot)$ in terms of its local utility functions alone. Just as in the case of functions of a single variable, once we've used the Fundamental Theorem of Calculus to exactly express the global change in a function in terms of the integral of its first derivatives, the inclusion of second and/or higher order derivatives can achieve no further increase in precision. Indeed, the suggestion of a higher order Taylor *approximation* to the preference functional would leave us in a *worse* position than my use of linear approximations, path integrals, and the Fundamental Theorem to achieve an exact representation of global changes in $V(\cdot)$.

3. Deriving Local Expected Utility Expansions for Russell's Insurance Problems

Russell argues his second assertion by presenting the following simple insurance problem: Consider a setting of two states of nature with probabilities p and $1-p$, where x_1 and x_2 denote the payoffs in the two states, and where the function $V(p,x_1,x_2)$ represents the individual's preferences over probability distributions. Specifically, he considers the case when:

> "we have $x_i = x_i(\beta)$, $i = 1,2$, where β is the choice parameter. Now $V(\beta) = V(p,x_1(\beta),x_2(\beta))$ can be expanded in a Taylor series whose first term is linear in x_i, but, in general this cannot be expected to be linear in p. There is no first order expected utility approximation in this case."

The conclusion that "there is no first order expected utility approximation" stems from a failure to recognize that the first order expansions in Machina (1982) (equations (2) and (3) above) are *not* with respect to *small changes in outcomes* ("a Taylor series whose first term is linear in x_i"), but rather, with respect to *small changes in cumulative distribution functions*. In order to represent this problem in terms of preferences over cumulative distribution functions and derive its local expected utility expansion, we define $\delta_x(\cdot) \in D[0,M]$ as the cumulative distribution function of the degenerate distribution that yields the outcome x with certainty.[3] The problem then becomes

$$\max_{\beta} V(p\cdot\delta_{x_1(\beta)}(\cdot) + (1-p)\cdot\delta_{x_2(\beta)}(\cdot)) \tag{6}$$

where the preference functional $V(\cdot)$ over cumulative distribution functions in $D[0,M]$ is assumed to be smooth in the sense of equation (2). Defining[4]

$$F_\beta(\cdot) \underset{\beta}{\equiv} p\cdot\delta_{x_1(\beta)}(\cdot) + (1-p)\cdot\delta_{x_2(\beta)}(\cdot) \tag{7}$$

and

$$V(\beta) \underset{\beta}{\equiv} V(p\cdot\delta_{x_1(\beta)}(\cdot) + (1-p)\cdot\delta_{x_2(\beta)}(\cdot)) \underset{\beta}{\equiv} V(F_\beta(\cdot)) \tag{8}$$

equation (3) then implies

$$\begin{aligned} V'(\beta_0) &= \left.\frac{dV(F_\beta)}{d\beta}\right|_{\beta=\beta_0} = \left.\frac{d}{d\beta}\Big(\int U(x;F_{\beta_0})\cdot dF_\beta(x) + o\big(\|F_\beta - F_{\beta_0}\|\big)\Big)\right|_{\beta=\beta_0} \\ &= \left.\frac{d}{d\beta}\big(p\cdot U(x_1(\beta);F_{\beta_0}) + (1-p)\cdot U(x_2(\beta);F_{\beta_0})\big)\right|_{\beta=\beta_0} + \left.\frac{d}{d\beta}\big(o\big(\|F_\beta - F_{\beta_0}\|\big)\big)\right|_{\beta=\beta_0} \\ &= p\cdot U'(x_1(\beta_0);F_{\beta_0})\cdot x_1'(\beta_0) + (1-p)\cdot U'(x_2(\beta_0);F_{\beta_0})\cdot x_2'(\beta_0) \end{aligned} \tag{9}$$

---

[3] Thus $\delta_x(\omega) \equiv 0$ for $\omega \in [0,x)$ and $\delta_x(\omega) \equiv 1$ for $\omega \in [x,M]$.

[4] I follow Russell in using $V(\cdot)$ to denote both the preference function(al) over distributions as well as the preference function over β it induces.

where $U'(x;F) = \partial U(x;F)/\partial x$.[5]

Thus, the first order expansion of $V(\cdot)$ (the derivative $V'(\beta_0)$) *does* possess a "local expected utility" structure, in the standard sense that it differs from the corresponding expected utility expansion *only* in the replacement of the von Neumann-Morgenstern utility function $U(\cdot)$ with the local utility function $U(\cdot;F)$ evaluated at the distribution $F(\cdot) = F_{\beta_0}(\cdot)$ about which the expansion is taken.[6] This contradicts the assertion that "There is no first order expected utility approximation in this case."

Russell gives a second example of an insurance problem in which the individual's choice may affect the support of the probability distribution. Once again, a proper application of the approach of Machina (1982) yields a first order expansion which is locally expected utility. This problem involves an individual with initial wealth w facing a random loss $\tilde{x}$ with cumulative distribution function $G(\cdot)$. Here, an insurance contract consists of a premium payment P and an "indemnity function" $I(\cdot)$ which represents the insurance payment received as a function of the amount of the loss. The random final wealth implied by an insurance contract $(P,I(\cdot))$ is thus

$$\tilde{y} \;=\; w - P - \tilde{x} + I(\tilde{x}) \tag{10}$$

Analogously to Russell's treatment of his first problem, let the premium $P(\beta)$ and the indemnity function $I(\cdot,\beta)$ depend on a choice parameter $\beta \in [\beta_{min},\beta_{max}]$, and for each β define $F_\beta(\cdot)$ to be the cumulative distribution function of the random variable[7]

$$\tilde{y}_\beta \;=\; w - P(\beta) - \tilde{x} + I(\tilde{x},\beta) \tag{11}$$

Once again, given a preference functional $V(\cdot)$ over $D[0,M]$ which is smooth in the sense of equation (2),[8] and defining

$$V(\beta) \;\underset{\beta}{\equiv}\; V(F_\beta(\cdot)) \tag{12}$$

equation (3) implies:

$$V'(\beta_0) \;=\; \left.\frac{dV(F_\beta)}{d\beta}\right|_{\beta=\beta_0} \;=\; \left.\frac{d}{d\beta}\left(\int U(y;F_{\beta_0})\cdot dF_\beta(y) + o\left(\left\|F_\beta - F_{\beta_0}\right\|\right)\right)\right|_{\beta=\beta_0} \tag{13}$$

---

5 $do(\|F_\beta - F_{\beta_0}\|)/d\beta|_{\beta=\beta_0} = 0$ follows since $d\|F_\beta - F_{\beta_0}\|/d\beta|_{\beta=\beta_0} = p\cdot|x_1'(\beta_0)| + (1-p)\cdot|x_2'(\beta_0)| < \infty$.

6 The expected utility version of this problem is $\max_\beta V(\beta)$, where $V(\beta) \equiv p\cdot U(x_1(\beta)) + (1-p)\cdot U(x_2(\beta))$. In this case $V'(\beta_0) = p\cdot U'(x_1(\beta_0))\cdot x_1'(\beta_0) + (1-p)\cdot U'(x_2(\beta_0))\cdot x_2'(\beta_0)$, which corresponds exactly to (9), with the local utility function $U(\cdot;F_{\beta_0})$ replaced by the von Neumann-Morgenstern utility function $U(\cdot)$.

7 For the following derivation, I assume that $\tilde{x}$, $P(\cdot)$, and $I(\cdot,\cdot)$ are bounded, so the supports of the random variables $\tilde{x}$ and $\{\tilde{y}_\beta | \beta \in [\beta_{min},\beta_{max}]\}$, *though not necessarily identical*, all lie within some bounded interval $[a,b]$, and also that the functions $P(\cdot)$, $I(\cdot,\cdot)$ and $U(\cdot;F_{\beta_0})$ are continuously differentiable. All integrals in equation (13) and in note 10 are thus over $[a,b]$.

8 Such as the ones in the above table.

$$= \frac{d}{d\beta}\left(\int U\left(w - P(\beta) - x + I(x,\beta); F_{\beta_0}\right) \cdot dG(x)\right)\Bigg|_{\beta=\beta_0} + \frac{d}{d\beta}\left(o\left(\left\|F_\beta - F_{\beta_0}\right\|\right)\right)\Bigg|_{\beta=\beta_0}$$

$$= \int U'\left(w - P(\beta_0) - x + I(x,\beta_0); F_{\beta_0}\right) \cdot \left[-P'(\beta_0) + I_2(x,\beta_0)\right] \cdot dG(x)$$

where $U'(x;F) = \partial U(x;F)/\partial x$ and $I_2(x,\beta) = \partial I(x,\beta)/\partial\beta$.[9] As before, this first order expansion differs from the corresponding expansion for pure expected utility preferences *only* in the replacement of the von Neumann-Morgenstern utility function $U(\cdot)$ with the local utility function $U(\cdot;F)$ evaluated at the distribution $F(\cdot) = F_{\beta_0}(\cdot)$ about which the expansion is taken.[10] Once again, nowhere do we need the assumption that the distributions $\{F_\beta(\cdot)\,|\,\beta \in [\beta_{min},\beta_{max}]\}$ or $G(\cdot)$ have identical supports.

4. References

Allen, B. (1987). "Smooth Preferences and the Local Expected Utility Hypothesis," *Journal of Economic Theory* 41, 340-355.

Apostol, T. (1974). *Mathematical Analysis, 2nd Ed.* Reading, Mass.: Addison-Wesley Co.

Bardsley, P. (1992). "Local Expected Utility Without Compactness," La Trobe University Economics Department Discussion Paper No. 28/92.

Bardsley, P. (1993). "Local Utility Functions," *Theory and Decision* 34, 109-118.

Billingsley, P. (1968). *Convergence of Probability Measures.* New York: John Wiley & Sons.

Chew, S. (1983). "A Generalization of the Quasilinear Mean With Applications to the Measurement of Income Inequality and Decision Theory Resolving the Allais Paradox," *Econometrica* 51, 1065-1092.

Chew, S., L. Epstein and U. Segal (1991). "Mixture Symmetry and Quadratic Utility," *Econometrica* 59, 139-163.

Chew, S., L. Epstein and I. Zilcha (1988). "A Correspondence Theorem Between Expected Utility and Smooth Utility," *Journal of Economic Theory* 46, 186-193.

---

9 The conditions of note 7 and Apostol (1974, p.167, Thm.7.40) allow us to differentiate under the Stieltjes integral sign in (13). They also imply that $d\|F_\beta - F_{\beta_0}\|/d\beta|_{\beta=\beta_0}$ is finite, and hence that $do(\|F_\beta - F_{\beta_0}\|)/d\beta|_{\beta=\beta_0} = 0$, whether or not changes in β affect the support of $F_\beta(\cdot)$.

10 The expected utility version is $\max_\beta V(\beta)$, where $V(\beta) \equiv \int U(w - P(\beta) - x + I(x;\beta)) \cdot dG(x)$. Thus $V'(\beta_0) = \int U'(w - P(\beta_0) - x + I(x,\beta_0)) \cdot [-P'(\beta_0) + I_2(x,\beta_0)] \cdot dG(x)$, which corresponds exactly to (13) with the local utility function $U(\cdot;F_{\beta_0})$ replaced by the von Neumann-Morgenstern utility function $U(\cdot)$.

Chew, S., E. Karni and Z. Safra (1987). "Risk Aversion in the Theory of Expected Utility with Rank Dependent Probabilities," *Journal of Economic Theory* 42, 370-381.

Chew, S. and N. Nishimura (1992). "Differentiability, Comparative Statics, and Non-Expected Utility Preferences," *Journal of Economic Theory* 56, 294-312.

Green, J. and B. Jullien (1988). "Ordinal Independence in Non-Linear Utility Theory," *Journal of Risk and Uncertainty* 1, 355-387.

Karni, E. (1987). "Generalized Expected Utility Analysis with State-Dependent Preferences," *International Economic Review* 28, 229-240.

Karni, E. (1989). "Generalized Expected Utility Analysis of Multivariate Risk Aversion," *International Economic Review* 30, 297-305.

Machina, M. (1982). " 'Expected Utility' Analysis Without the Independence Axiom," *Econometrica* 50, 277-323.

Machina, M. (1984). "Temporal Risk and the Nature of Induced Preferences," *Journal of Economic Theory* 33, 199-231.

Machina, M. and W. Neilson (1987). "The Ross Measure of Risk Aversion: Strengthening and Extension," *Econometrica* 55, 1139-1149.

Röell, A. (1986). "Portfolio Choice in Quiggin and Yaari's Rank-Order Model of Choice Under Uncertainty," manuscript, London School of Economics.

Russell, T. (1996). "Generalised Expected Utility and the Demand for Insurance: The Limits of Machina," in this volume.

Wang, T. (1993). "L_p-Fréchet Differentiable Preference and 'Local Utility' Analysis," *Journal of Economic Theory* 61, 139-159.

NECESSARY CONDITIONS FOR EFFICIENT MULTIPLE-BID AUCTIONS

FRIEDEL BOLLE*
Europa-Universität Viadrina
Frankfurt (Oder)
Große Scharrnstraße 59
15230 Frankfurt (Oder)

Abstract:

m identical objects are to be sold in a multiple-bid auction where the m highest bids are accepted. An efficient private value auction must induce an identical transformation of all values of the bidders into bids. (In a common value auction the problem of allocative efficiency disappears.) The pricing rule has to be an additively separable function of the bids.Under the condition that a pricing rule is efficient, monotone, and reciprocitive (a bidder has to pay only for the units received) it has to obey certain restrictions. These restrictions are not met by the pricing rules found in large-scale auctions as the British electricity auction and the auctions organized by the American Treasury Department and by the German Bundesbank. If the pricing rule is "simple" the Vickrey rule and trivial transformations of the Vickrey rule are the only efficient pricing rules.

* I would like to thank Werner Güth for helpful comments on a previous version of this paper.

R. Nau et al. (eds.), Economic and Environmental Risk and Uncertainty, 123–143.

1 Introduction

In multiple-bid auctions a number of identical objects are to be sold (or bought). The potential buyers (sellers) submit offers for all or some of the objects. These offers need not to be equal. According to the rules of the auction, successful offers are selected and prices for these offers are determined. There are some examples of large-scale multiple-bid auctions, the most prominent of which are the British electricity auction, the auction of Treasury Bills in the United States (and perhaps in other countries), and the auction to provide liquidity to the banking sector (Wertpapierpensionsgeschäfte) by the German Bundesbank. In all three cases, the respective transaction volume runs into billions of dollars per week. In addition to these large-scale examples, there are medium-scale auctions, such as gold auctions and the proposed auctions of CO_2 or SO_2 emission rights. Furthermore, there are small-scale mutiple-bid auctions for public work, used army materials, and others.

Bidding behaviour and market entry depend on the rules of the auction, the most important of which is the determination of prices which successful bidders have to pay. In the cases of the auctions of American Treasury Bills and Bonds, as well as in the case of the German Wertpapierpensionsgeschäfte, the choice of the appropriate pricing rule has been discussed publicly (see The Economist, September 26th, 1992, p. 91, and Die Deutsche Bundesbank, 1989). In both auctions the pricing rules have been changed in the course of time. In the British electricity market, the efficiency of the massive market entry of Combined Cycle Gas Turbines (CCGT) has been discussed in connection with the crisis of British Coal (see Power in Europe, November 6th, 1992, pp. 2-6). Although there may be many reasons for inefficient market entry, one precondition to prevent it is to secure an efficient procedure of the auction is the efficiency of the auction procedure.

The result of an auction is efficient if the producers with the lowest costs deliver the objects concerned, or, vice versa, if the buyers with the largest willingness to pay receive the objects. From a social point of view, auction procedures with efficient outcomes, i.e. efficient auctions, should be used. In this paper, the auction procedures used in the three large-scale auctions turn out to be inefficient.

In the next section, the efficiency problem is explained by means of an example. In sections III and V, necessary conditions for efficient auctions are derived. It is shown in section VI that these conditions are not fulfilled in the large-scale examples of multiple bid auctions. In section IV the (efficient) Vickrey auction is described and in VIII we discuss whether this auction should be used. In section VII multiple-bid auctions and competition with supply functions are compared.

The formalization of the problem is close to Engelbrecht-Wiggans (1988).

2 The efficiency problem

Let me illustrate the efficiency problem which is inherent in multiple-bid auctions by using a simple example. There is an owner of three identical objects and there are two potential buyers. The latters' valuations of the objects (marginal willingness to pay) are $x^l = (x_1^l, x_2^l, x_3^l)$, i.e. the buyers evaluate the possession of one object with x_1^l, the possession of two objects with $x_1^i + x_2^i$, and the possession of three objects with $x_1^i + x_2^i + x_3^i$. Buyer 1 knows x^1 and assumes x^2 to be independent of x^1 and to take the values $x^2 = (3, 2, 0)$ or $x^2 = (3, 1, 0)$ or $x^2 = (0, 0, 0)$ with a probability 1/3 each. The same applies for buyer 2. So they are in a symmetric situation.

The seller announces that he will allocate the objects to the three highest bids at the price of the third highest bid. (This pricing rule is used in the British electricity auctions as well as in the auction of bonds by the American Treasury up to 1992.) In case of ties, the winning bid is selected at random. The seller sets a reservation price of 0, i.e. no bid below 0 will be accepted.

The symmetric auction game thus defined has no symmetric Nash equilibria in pure strategies. I will show this in two steps. Without loss of generality we can assume the bids to be ordered, $b_1(x^i) \geq b_2(x^i) \geq b_3(x^i)$. $b_2(3, x, 0) < x$, $x = 1, 2$, cannot belong to a symmetric equilibrium because both buyers could increase their bid by a small amount thus winning a second object with probability 1 and thus increasing their expected profit. $b_2(3, x, 0) > x$ cannot belong to a symmetric equilibrium either, because there is a positive probability to get the object at the price b_2 which is larger than the value of the object. Because of the same argument we have to conclude that $b_3(3, x, 0) \leq 0$ and $b_1(0, 0, 0) \leq 0$.

Symmetric strategies with these restrictions, especially $b_2(3, x, 0) = x$, imply unique profits.

$$G_i = \begin{cases} 3+x & \text{if } x^i = (3,x,0) \text{ and } x^j = (0,0,0) \\ 2 & \text{if } x^i = x^j = (3,1,0) \\ 1 & \text{if } x^i = (3,x,0) \text{ and } x^j = (3,2,0) \\ 0 & \text{otherwise} \end{cases}$$

If i reduces his bids in the case $x^i = (3, 2, 0)$, say he bids $b_2 = 1.5$, then the above profits remain unchanged except for the cases $x^i = (3, 1, 0)$ where G_i increases.

So there is no symmetric Nash equilibrium in pure strategies. There are, however, asymmetric Nash equilibria in pure strategies, namely

$$b^1(x^1) = \begin{cases} (3, x, -1) & \text{for } x^1 = (3,1,0) \\ (-1,-1,-1) & \text{for } x^1 = (0,0,0) \end{cases}$$

$$b^2(x^2) = (0,0,0)$$ [1]

and vice versa. These two equilibria have similarities with the "battle of sexes" equilibria. The buyer who bids (0, 0, 0) earns the lower expected profit.

---

1 This is an equilibrium because it does not pay for bidder 2 to fight for a second object in cases $x^j = (3, x, 0)$. He would increases the price of all objects from 0 to more than 1.

The problem with asymmetric equilibria - as well as with mixed strategy equilibria - is that the values are transformed into bids by means of different bidding functions. Thus the order of bids needs no longer reflect the order of values. In the above asymmetric equilibria, a bidder with a value $x_2^i = 1$ may win a second object though the other's value is $x_2^j = 2$. So there is a positive probability of an inefficient allocation.

What is responsible for the inefficiency result above? It is the design of the auction, in particular the choice of the pricing rule. In the following sections of the paper we will investigate more general auction situations and classes of pricing rules. It will turn out that, under certain conditions, most pricing rules suffer from the possibility of inefficient equilibrium allocations.

Intuition is often guided by auctions with identically distributed independent private values and single bids[2]. In such cases, efficiency is no problem and thus, it has not aroused much interest[3]. Efficiency in multiple-bid auctions has been investigated by Dubey and Shubik (1980), in a model with double auctions and complete information, however. Often authors have focused on optimal auction design (Myerson, 1981; Bulow and Roberts, 1989; Mascin and Riley, 1989; Branco, 1992; and many others), i.e. the question of how an auctioneer can sell an object (several objects) at the highest price or buy an object at the lowest price. While this can be expected to be the goal of a private seller the main goal of a public auctioneer, a Treasury Department or a Federal Reserve Bank, should be efficiency[4].

It is well-known that efficient equilibria result from the Vickrey rule[5] (Vickrey, 1961, for a formal definition see Section IV) where truth revealing is a dominant strategy for every bidder. In the example above, this would result in expected payments of 1 to the seller while the asymmetric equilibria under the above pricing rule imply expected payments of 0. The Revenue Equivalence Theorem[6] (Myerson,1981; Engelbrecht-

---

2 Although there may be many objects, each buyer submits only one bid.

3 There are papers, for example Harris and Raviv (1981), which investigate efficiency explicitly. These investigations, however, are restricted to single-bid auctions.

4 There need not be but often there will be a contradiction between the goals of optimal auction design (revenue maximization of the auctioneer) and efficient auction design (optimal allocation). If the auction is meant to raise income for the government and be a substitute for taxes, the efficiency problem has to be discussed in a broader context. That is beyond the scope of this paper.

5 In a multiple-bid Vickrey auction the m highest bids are selected. A bidder who wins m_i units has to pay the sum of the m_i highest bids which are not selected and which are not his own bids. A Vickrey auction with m = 1 is called a second price auction.

6 The Revenue Equivalence Theorem states that, under certain conditions, a bidder with a given vector of values (a vector of production costs) has to pay (will be

Wiggans,1988) cannot be applied. In Bolle (1993) it is shown more generally that the applicability of this Theorem is rather restricted in the case of multiple-bid auctions.

3 Necessary conditions for efficient auctions

In the following, we will investigate sell auctions like the auctions of the American Treasury Department. Of course, after applying an appropriate transformation, this approach can be used also for buy auctions.

Assumption 1 (Independent private values and priors with positive density function): Each bidder evaluates the m identical units which a seller offers by a vector[7] $\mathbf{x}^i = (x^i_1,...,x^i_m)$ of money values with $x^i_j \geq x^i_{j+1} \geq 0$. The x^i_j can be regarded as the incremental values $x^i_j = U^i(j) - U^i(j-1)$ of a utility function $U^i(j)$ for the possession of j units. Each bidder knows his own valuation but not the valuation of the other bidders. i believes that the $\mathbf{x}^k$, $k \neq i$, are selected according to independent random distributions with distribution functions $F^k(\mathbf{x}^k)$ on $[0,1]^m$. The F^ks are based on continuous density functions $f^k(\mathbf{x}^k)$

$$F^k(x^k) = \int_0^{x_1^k} \cdots \int_0^{x_m^k} f(\xi^k{}_1,...,\xi^k{}_m) d\xi^k{}_1 d\xi^k{}_m, \tag{1}$$

with

$$f^k(\mathbf{x}^k) \begin{cases} > 0 \text{ for } \mathbf{x}^k \text{ with } x^k_j > x^k_{j+1} \text{ for all j,} \\ = 0 \text{ for } \mathbf{x}^k \text{ with } x^k_j < x^k_{j+1} \text{ for one j.} \end{cases}$$

Note that bidders' values are independent of the bidders' wealth (i.e. independent of what they have to pay for subsequent units).

Assumption 1 introduces private values. This restriction is less severe than it seems to be. In the other extreme case, namely when common values are introduced, the efficiency problem disappears because it does not matter who is going to win the objects. At least the British electricity auction is a pure private value auction. (In this sell auction, the values are the costs of the producers.) The cases of the other two large scale auctions are discussed in Section VI.

Assumption 2 (Risk-neutrality): All bidders are risk-neutral.

---

paid) the same expected aggregate price under any pricing rule. Then the revenue of the seller of the m objects (the aggregate costs of the public authority in our example) is independent of the pricing rule of the auction.

7 In the following, vectors are indicated by bold type.

The bidders are required to submit independent bids for all units. $b^i(\mathbf{x}^i)=(b^i_j(x^i))_{j=1,...,m}$ is the vector of bids of the ith bidder. Let us restrict our attention to the class of auctions for which the following allocation rule is used.

Assumption 3 (Allocation Rule): The seller sets a reservation price of 0. The m highest bids (≥ 0) of all bidders are selected[8]. In case of ties a random selection mechanism is used.

We will see that we need not specify this random mechanism in the case of efficient bidding equilibria because ties occur with probability 0 (because of Assumption 1 and Proposition 1). In the following we will concentrate on such efficient equilibria.

Definition: A Nash equilibrium of bidding strategies $b^i(x^i)$ is called an efficient bidding equilibrium if the bids connected with the m highest values x^i_j are selected with probability 1. An auction with an efficient bidding equilibrium is called an efficient auction.

Let $b^i(\mathbf{x}^i) = (b^i_1(\mathbf{x}^i),..., b^i_m(\mathbf{x}^i))$, $i = 1,..., n$ constitute a Nash equilibrium of bidding strategies. Intuitively, it is clear that all bidding strategies have to be "simple" and equal (see (2) below). Otherwise, the bids will not reflect the true order of the values.

Proposition 1: In an efficient bidding equilibrium, all bidding strategies $b^i_j(\cdot)$ have to be equal. There is a strictly monotone function g so that

$$b^i_j(\mathbf{x}^i) = g(x^i_j) \tag{2}$$

is fulfilled almost everywhere for all $i = 1,..., n$, $j = 1,..., m$.

Proof: See Appendix.

If (2) cannot be fulfilled either because there is no equilibrium at all or because there are only equilibria with different bidding strategies, then the pricing rule causes an inefficient auction. What we want to do is to derive *necessary* conditions for efficient auctions. Thus we have to proceed with

Assumption 4 (Efficiency): The auction results in an efficient equilibrium, i.e. (2) is fulfilled.

In the following, we will ask under which pricing rules of an auction (2) can be fulfilled. We will proceed in a similar way as Engelbrecht-Wiggans (1988). Let us also follow Engelbrecht-Wiggans in *suppressing the index i* (in most cases).

Let $p_j(\mathbf{b},\mathbf{w})$ denote the probability[9] that bidder i will receive at least j units when he bids **b** and when the others bid **w**. **w** is a vector (not a matrix) with m(n-1) components which contains all the above b^k_l, $k \neq i$ in an ordered sequence, i.e. $w_j \geq w_{j+1}$.

---

8 This Assumption can be substituted by an assumption on envyfreeness (see Güth, 1986).

9 Under Assumption 3, in connection with Assumption 1 and $g'>0$ (Proposition 2), p_j is different from 0 and 1 only in the case of ties, i.e. on a set of measure 0.

Because of Assumption 3, p_j depends only on b_j and **w**. Let $P_j(\mathbf{b})$ denote the probability that bidder i will receive at least j units when he bids **b**. The bid b_j is successful if and only if it is higher than w_{m-j+1}. So P_j is completely determined by the distribution F_{m-j+1} of w_{m-j+1}, the (m-j+1)th of the ordered bids of the others. Thus we get two representations of P_j.

$$P_j(\mathbf{b}) = P_j(b_j) = \int p_j(bj, \mathbf{w})\, dF(\mathbf{w}) \tag{3}$$

where F(**w**) is the distribution function of **w**[10].

$$P_j(b_j) = F_{m-j+1}(b_j). \tag{4}$$

The payments of bidder i (the pricing rule) are denoted by c(**b**,**w**). The expected payments are

$$C(\mathbf{b}) = \int c(\mathbf{b},\mathbf{w})\, dF(\mathbf{w}). \tag{5}$$

Thus the expected utility of i is

$$U = \sum_{j=1}^{m} x_j\, P_j(b_j) - C(\mathbf{b}) \tag{6}$$

(5) requires c(**b**,**w**) to be measurable with respect to **w**; but we need more than this.

Assumption 5 (Pricing Rule): c(**b**,**w**) is measurable with respect to **w** and is differentiable with respect to **b** almost everywhere.

Up to now, we have not restricted the set of feasible bidding functions in any respect. Let us now require

Assumption 6 (Feasibility): A bidding function g(.) is feasible if it is differentiable almost everywhere[11].

Assumption 6 implies that, if there are values for which g(.) is non-differentiable then we also find an isolated point of non-differentiability. In the Appendix, we will see that such cases are incompatible with the monotonicity of g(.). So we get

Lemma 1: U is differentiable with respect to **b** for $\mathbf{b} = (g(x_1),\ldots, g(x_m))$. Therefore $b_j = g(x_j)$ is a solution of each of the equations

$$\partial U / \partial b_j = x_j \frac{\partial P_j(b_j)}{\partial b_j} - \frac{\partial C(b)}{\partial b_j} = 0, \quad j = 1,\ldots, m. \tag{7}$$

Proof: See Appendix

---

10 The distributions F^j of Assumption 1 are connected with the distributions $F_k(w_k)$ and F(w) via the bidding transformation g and the ordering process.

11 This assumption is sufficient but possibly not necessary. For an example where the differentiability of the bidding functions is derived from the requirement of measurability see Plum (1992).

Proposition 1 implies that $x_j = g^{-1}(b_j)$ is well defined. So (7) can be rewritten as

$$g^{-1}(b_j)\frac{\partial P_j(b_j)}{\partial b_j} = \frac{\partial C(b)}{\partial b_j} \tag{8}$$

for j = 1,..., m and all **b** $[g(0), g(1)]^m$.

Intregrating (7) we find

$$C(b) = \sum_{j=1}^{m} C_j(b_j) \tag{9}$$

with

$$C_j(b_j) = C_j^0 + \int_{\beta = g(0)}^{b_j} g^{-1}(\beta)\frac{dP_j(\beta)}{d\beta}\, d\beta . \tag{10}$$

Thus we get

Proposition 2: For efficient auctions, the bidding strategy g(.) has a positive derivative. The expected payments C(**b**) of a bid **b** have to be an additively separable function of the b_j.

$$\frac{dC_j(\beta)/d\beta}{dP_j(\beta)/d\beta} = g^{-1}(\beta) \tag{11}$$

is independent of j.

Proposition 2 poses restrictions on efficient pricing rules. On the other hand, (11) shows that there are expected payments $C_j(b_j)$ for all distributions F (or P_j, respectively) and all monotone bidding rules g which guarantee an efficient outcome. It is also easy to connect these expected payment functions with pricing rules c(**b**,**w**). However, such pricing rules are efficient only in connection with certain distributions F(**w**), and they are, possibly, rather complicated while existing pricing rules are "simple".

Assumptions 1 to 6 constitute a set of sufficient conditions for the Revenue Equivalence Theorem (Myerson, 1981; Engelbrecht-Wiggans, 1988, for multiple-bid auctions). (9) and (10) provide us with the Revenue Equivalence Theorem in our context, i.e. under our assumptions. After substituting b_j by $g(x_j)$ and $g^{-1}(\beta)$ by the value ζ, $C_j^*(x_j) = C_j(g(x_j))$ depends only on $dP_j(g(\zeta))/d\zeta$,. As g(.) is the same for all bidders and all bids, $dP_j/d\zeta$, is the same as in the case where everybody bids his true values. Thus, (10) implies that the expected payment under all efficient auctions should be equal (if, for example, $c_j^*(0)$ is required to be 0 in every case). In Bolle (1993), however, it is shown that the Revenue Equivalence Theorem cannot apply for a class of auctions to which our three large scale examples also belong. The arguments are similar as in the proof of Proposition 3 below. Before we proceed with extending Proposition 2 to the level of pricing rules c(**b**,**w**) let us first examine an example of a generally efficient pricing rule, namely the Vickrey Rule.

4 The Vickrey Rule

In a multiple-bid Vickrey Auction the m highest bids are selected. A bidder who wins m_i units has to pay the sum of the m_i highest bids which are not selected and which are not his own bids. This rule was first suggested by Vickrey (1961, p. 10) who explains it for the case of continuous bids, i.e. supply or demand functions. The above formulation can be found in Weber (1983).

$$c^v(\mathbf{b},\mathbf{w}) = \sum_{j=1}^{m} w_{m-j+1}\, I(b_j > w_{m-j+1}) \tag{12}$$

with

$$I(b_j > w_{m-j+1}) = \begin{cases} 1 \text{ if } b_j > w_{m-j+1} \\ 0 \text{ otherwise} \end{cases}. \tag{13}$$

In order to confirm that (12) and (13) coincide with the verbal explanation above let us start with j = 1. If b_1, i.e. the bidder's highest offer, is larger than w_m, i.e. the other bidders' m^{th} highest offer, then b_1 is accepted and w_m has to be paid. Now take j = 2. If b_2 is larger than w_{m-1} then b_2 is also accepted and w_{m-1} has to be paid by the successful bidder. This is repeated until b_k is smaller than w_{m-k+1}. Then b_k and all following (smaller) bids are not accepted.

We already know (see Vickrey, 1961) that truth revealing

$$b_j = x_j, \quad j = 1,...m \tag{14}$$

is a dominant strategy in the Vickrey Auction and that, under this strategy, an efficient allocation results. We can, however, derive (14) also from the necessary conditions of the last section. (9) and (12) imply

$$C^v(\mathbf{b}) = \int c^v(\mathbf{b},\mathbf{w})\, dF(\mathbf{w}) \tag{15}$$

$$= \sum_{j=1}^{m} \int_{w_{m-j+1}=g(0)}^{b_j} w_{m-j+1}\, dP_j(w_{m-j+1})$$

P_j is the distribution function of w_{m-j+1} (see 3b).

Thus we get

$$\partial C^v(\mathbf{b}) / \partial b_j = b_j \frac{dP_j}{db_j} \tag{16}$$

and from (8) follows

$$x_j = g^{-1}(b_j) = b_j \text{ for all j.} \tag{17}$$

There is a class of (trivial) efficient pricing rules connected with the Vickrey Auction, namely all

$$c^{vh}(\mathbf{b},\mathbf{w}) = c^v(h(\mathbf{b}), h(\mathbf{w})) \tag{18}$$

with $h(\mathbf{b}) = (h(b_1),..., h(b_m))$ and $h' > 0$.

Apparently, (16) and (17) are fulfilled with $h(b_j)$ instead of b_j, i.e. $g^{-1} = h$. Under this pricing rule, the Vickrey Rule is applied to the transformed bids $h(b_j)$.

5 Necessary conditions for the pricing rule

In the following, we will restrict our investigation to a smaller set of pricing rules. The restrictions are, however, fulfilled for most pricing rules used in practice.

Definition: If $c(\mathbf{b'},\mathbf{w}) \geq c(\mathbf{b},\mathbf{w})$ for $b'_j \geq b_j$ (all j) and for all $\mathbf{w}$, then $c(\mathbf{b},\mathbf{w})$ is called a *monotone* pricing rule.

Definition: $c(\mathbf{b},\mathbf{w})$ is called a *reciprocitive* rule if a bidder has to pay only for those units which he receives. Formally, this requires that there are $c_j(\mathbf{b},\mathbf{w})$ with

$$c(\mathbf{b},\mathbf{w}) = \sum_{j=1}^{m} c_j(\mathbf{b},\mathbf{w})\, I\,(w_{m-j+1} < b_j). \tag{19}$$

(19) is not fulfilled for "dollar auctions" (see Shubik, 1971; Leininger, 1989) which are used to model patent races or arms races. In these games, the "bids" are money spent (c.f. research expenditures) which is lost whether or not the "bidder" wins a unit (a patent).

Proposition 3: An efficient, monotone and reciprocitive pricing rule almost everywhere has the form

$$c(\mathbf{b},\mathbf{w}) = \sum_{j=1}^{m} c_j(b_j,\mathbf{w})\, I(w_{m-j+1} < b_j), \tag{20}$$

i.e. c_j does not depend on b_k, $k \neq j$.

If $c_j(b_j,\mathbf{w}) \geq 0$ and if $c_j\,(b_j, \mathbf{w})$ is differentiable then

$$c_j'(g(1),\mathbf{w}) = 0,\ j = 1,\dots, m-1 \tag{21}$$

with $c_j' = \partial c_j/\partial b_j$ is fulfilled almost everywhere.

Proof: See Appendix

The first part of Proposition 3 is rather plausible because of (9). The proof, however, is a little tedious. In the following, I want to make plausible the fact that the Vickrey Auction is the only linear efficient auction.

Definition: A monotone reciprocitive pricing rule is called *linear* if the c_j from (19) have the form

$$c_j(\mathbf{b},\mathbf{w}) = \sum_{k=1}^{(n-1)m} \alpha_{jk} w_k\,. \tag{22}$$

Note that, because of (20), c_j cannot linearly depend on b_k, $k \neq j$. Because of (21), c_j cannot linearly depend on b_j either.

Let us define the conditional expectation[12]

$$E(c_j(b_j, w) / w_{m-j+1} = b_j) = \frac{C_j'}{P_j'}. \tag{23}$$

From Proposition 2 follows

$$\frac{C_j'(\beta)}{P_j'(\beta)} = \sum_k \alpha_{jk} E(w_k / w_{m-j+1} = \beta) = g^{-1}(\beta). \tag{24}$$

Because of

$$E(w_k / w_k = \beta) = \beta, \tag{25}$$

we get

$$\begin{aligned} \frac{C_j'(\beta)}{P_j'(\beta)} - \frac{C_m'(\beta)}{P_m'(\beta)} &= (\alpha_{j,m-j+1} - \alpha_{m,1})\beta \\ &+ \sum_{k \neq j} \alpha_{jk} E(w_k / w_{m-j+1} = \beta) \\ &- \sum_{k \neq 1} \alpha_{mk} E(w_k / w_1 = \beta) \\ &= 0, \qquad j = 1,\ldots, m-1. \end{aligned} \tag{26}$$

The direct implication of (26) can be formulated as the following proposition.

Proposition 4: If the 2m-1 functions $E(w_k/w_{m-j+1} = \beta)$, $k \neq m-j+1$, $E(w_k/w_1 = \beta)$, $k \neq 1$, and $E(w_k/w_k = \beta)$ are linearly independent for every j, then a linear efficient pricing rule is equal to the Vickrey Rule multiplied by a constant factor.

Proof: From the linear independence of the functions $E(w_k/w_{m-j+1} = \beta)$ it follows that all coefficients in (26) have to be 0, i.e. the pricing rule is the Vickrey Rule multiplied by α_{m1}.

The requirement that the $E(w_k/w_{m-j+1} = \beta)$ are linearly independent is a rather weak condition. Only special distributions of the values of the bidders imply dependencies. As an example let us consider the case n = m = 2.

According to Proposition 4, we have to investigate the linear dependency of the three functions

$$E(w_k / w_k = \beta) = \beta \tag{27}$$

$$E(w_2 / w_1 = \beta) = \left(\int_0^\beta w_2 f(\beta, w_2) dw_2 \right) / \left(\int_0^\beta f(\beta, w_2) dw_2 \right) \tag{28}$$

---

12 E is merely an "abbreviation" of the right hand side of (23). We do not utilize the fact that E is the conditional expectation of $c_j(b_j,\mathbf{w})$ under the condition that $w_{m-j+1}=b_j$. For the derivation of C_j' and P_j' see (A6) and (A8) in the appendix.

$$E(w_1 / w_2 = \beta) = \left(\int_{\beta}^{1} w_1 f(w_1, \beta) dw_1 \right) / \left(\int_{\beta}^{1} f(w_1, \beta) dw_1 \right) \qquad (29)$$

Let us concentrate on polynomial density functions[13]. For $f(w_1, w_2) = a\ w_1^r\ w_2^s$ we get

$$E(w_2 / w_1 = \beta) = \frac{\beta^r \beta^{s+2} /(s+2)}{\beta^r \beta^{s+1} /(s+1)} = \frac{s+1}{s+2} \beta \qquad (30)$$

and therefore a linear dependency. For all other polynomial density functions, cf. $f(w_1,w_2) = (aw_1^q + bw_1^r)(cw_2^s + dw_2^t)$, the three conditional expectations are linearly independent.

So a linear pricing rule different from the Vickrey Rule is efficient only in exceptional cases. In particular, it is not efficient *for all distributions* of values.

Let us now evaluate the most important examples of multiple-bid auctions.

6 Extant large-scale multiple-bid auctions

Since April 1, 1990, the electricity industry in England and Wales has been given a competitive structure. Every day the electricity producers submit bids for the following day's spot market. As a first approximation, they require an average price for each of their generators. Because of changing demand, every half hour a new market-clearing price is determined by the lowest excepted bid[14]. For further details[15], see Green (1991) and Capel & Co (1990).

The British electricity auctions are *private value auctions* where the values are determined by the average production costs of the generators. The difference to the formal model above is that the same bid is used for 48 auctions with different quantities. Under the assumed risk-neutrality, however, this does not seem to have an essential influence on the results.

---

13 Every density function $f(w_1, w_2)$ can be approximated by a sequence of polynomial functions.

14 The respective auction is often called a Dutch auction or a single-price auction.

15 Several additional rules, for example with respect to ties or with respect to limited transmission capacity in the case of British electricity auctions, are applied in the three large-scale auctions. By neglecting these rules the analysis is considerably simplified.

Of similar importance are auctions within the framework of the "open market policy" of the German Bundesbank which also used this rule up to 1988. In both cases, the transaction volumes run into billions of dollars per week.[16]

Again, this auction is, for the largest part, a private value auction. The use and the value of the additional liquidity which the Bundesbank provides is predetermined by the participating banks. Only a small amount of this liquidity is redistributed on a secondary market. The private value motivation of the banks is also reflected in the extremely variable participation rates. For further arguments see Nautz (1994).

The formal pricing rule of these auctions is that of the introductory example, namely

$$c(\mathbf{b},\mathbf{w}) = \sum_{j=1}^{m-1} c_j(b,w) I(w_{m-j+1} < b_j) \tag{31}$$

with

$$\begin{aligned} c_j(\mathbf{b},\mathbf{w}) = {} & b_j I(w_{m-j} > b_j) \\ & + w_{m-j} I(w_{m-j} < b_j) I(w_{m-j} > b_{j+1}) \\ & + b_{j+1} I(w_{m-j} < b_{j+1}) I(w_{m-j-1} > b_{j+2}) \\ & + \ldots \end{aligned} \tag{32}$$

As (32) shows $c_j(\mathbf{b},\mathbf{w})$ is not independent of b_k, $k \neq j$, as required by Proposition 3.

Since 1988, the German Bundesbank used another pricing rule: a successful bidder has to pay according to his bid. This is the same rule as that used, up to September 1992, by the American Treasury for the auctioning of bonds.

The latter auction and similar auctions in other countries (e.g. Austria) have more the character of a common value auction because a large share of the bonds are resold on a secondary market. For those bidders, however, who are firmly determined to keep there bonds and for the primary dealers who buy on the behalf of customers values are private. The fact that, for primary dealers with orders to buy at certain prices, strategic bidding is profitable (implied by Proposition 3) was the basis for the Salomon Brothers scandal and, at last, the reason for the change of the pricing rule in September, 1992.

To pay according to one´s bids means

$$c(\mathbf{b},\mathbf{w}) = \sum_{j=1}^{m} b_j I(w_{m-j+1} < b_j). \tag{33}$$

From Proposition 3, Equation (21), it follows that this pricing rule is inefficient.

Thus, the pricing rules of all three examples of large-scale multiple-bid auctions have turned out to be inefficient for private value auctions. (As already mentioned, for common value auctions no problem of allocative efficiency exists.) The American Trea-

---

16 Contrary to the American sell auctions of Treasury Bills, the German Bundesbank buys and sells back (after a predetermined time interval) certain bonds in order to provide the banking sector with liquidity.

sury and the German Bundesbank have changed their pricing rules during the past few years. However, they only switched over from one inefficient rule to another.

It is easy to get the same inefficiency result for variants of the above rules. "Simple" rules are shown to be inefficient because of Proposition 4. "Complicated" rules usually do not show the structure required in Proposition 3. If we determine, for example, a single price by the best rejected bid or by the best rejected bid of other bidders, Proposition 3 characterizes these rules as inefficient. (The formalization of the pricing rules is left to the reader.) Proposition 4 excludes such simple alternatives as $c_j=w_k$, $k \neq m-j+1$, or the determination of c_j as an average of some bids.

7 Competition with demand or supply functions

If the number of units approaches infinity (the size of the units approaches 0), the valuation vector of a bidder may be turned into a utility function U(y,a), $y \in [0,1]$ where U(y,a) denotes the aggregate utility for a share of y of the total number of objects. $U' > 0$, $U'' < 0$ is assumed. The joint distribution function for the valuation of the m objects is replaced by a probability measure on a set A which is an index set for the utility functions U(y,a), i.e. $a \in A$. The vector of bids is replaced by a bidding function b(y) where $db/dy = b'(y) > 0$ are the bids for the "marginal units". Because these bids are ordered, $b''(y) \leq 0$ results.

If a single price p is determined by the pricing rule, this model can be viewed as competition with demand functions $Y=b'^{-1}(p)$. A similar case ("counterspeculation") of competition with demand and supply functions has been investigated by Vickrey (1961). Only recently have supply function equilibria been investigated under somewhat different condititions (Wilson, 1979; Klemperer and Meyer, 1989; Bolle, 1991, 1992; Green and Newbery, 1991; Newbery, 1991). In the previous sections, bidders do not know the other bidders' utility functions but are perfectly informed about the demand or the supply. In the recent investigations of supply function equilibria, bidders have complete information about the others' valuations (cost functions), but do not know the size of demand[17].

Except for an extreme case with respect to the distribution of demand (see Klemperer and Meyer, 1989), the above inefficiency results (on uniform-price auction rules) are confirmed. Usually there are multiple equilibria. In *symmetric* models marginal cost supply functions constititute an efficient equilibrium, but it is the most unattractive one from the point of view of the bidders. All bidders can be better off by switching to other equilibria. There is one equilibrium which allows all of them the largest profit. It seems to be likely that this specific equilibrium is chosen without the necessity of precommunication.

---

17 There is also a "hybrid" auction model of von der Fehr and Harbord (1991) where there is incomplete information on the valuations of the other bidders as well as stochastic demand. Inefficient allocation is found in this model, too.

8 Should we use the Vickrey Auction?

The Vickrey Rule has turned out to be the only "simple" efficient multiple-bid pricing rule (except for trivial transformations). Remember that, under this rule, a bidder who wins m_i units (m_i of his bids belong to the m highest bids) has to pay the sum of the m_i highest bids which are not selected and which are not his own bids. Such a rule seems to be easy to implement in a sealed bid auction[18].

In spite of this, there does not seem to be any important example of a multiple-bid auction where this rule is used. The key to understanding this phenomenon may lie

- in the unwillingness of the bidders to uncover their true valuation if competitors or the auctioneer could take advantage of this either in the short or the long run (Rothkopf et al.,1990);
- in the larger vulnerability of the Vickrey Rule to collusive behavior (Robinson, 1985; Ungern-Sternberg, 1988; Güth and van Damme, 1986).

In single-unit second price auctions, the following strategies

$$b^i{}_1 = \begin{cases} 1 \ for\ i = k \\ 0 \ for\ i \neq k \end{cases} \tag{34}$$

constitute a Nash equilibrium where bidder k gets the object with certainty at a price of 0. (For the simple proof, remember assumption 1.) The question, of course, is why the bidders $i \neq k$ should be so "altruistic". Scherer (1988) emphasizes that collusion in such a case would work only if the $i \neq k$ are paid directly by side payments or indirectly by a believable promise that he will be the winner in a subsequent auction.

Cooperation is less difficult in multiple-bid auctions. The bidding strategies

$$b^i{}_1 = \begin{cases} 1 \ for\ j = 1, ..., m_i;\ i = 1, ..., m \\ 0 \ otherwise \end{cases} \tag{35}$$

with $$\sum_{i=1}^{n} m_i = m, \quad m_i \geq 0,$$

constitute Nash equilibria.

If m is large compared to n (which is the case in all of the important multiple-bid auctions) then neither side-payments nor long-term agreements among the bidders are

---

18 Provided that there are enough rejected bids. In the British electricity market, for example, there may be more accepted bids of National Power, the largest competitor, than rejected bids of other firms. This is a pathological case under any pricing rule. If demand is predetermined, some bids of National Power have to be accepted. In this (but also in the general) case, the Vickrey Rule can be modified by introducing a reservation price (= arbitrarily many hypothetical bids with this price). In the electricity market, we can supplement these pricing rules with an obligation to produce the quantity demanded.

necessary. A "fair" distribution of profits is possible in the one-shot game. The m_i might even be reached in a process of tacit collusion.

So, in multiple-bid auctions, we seem to be placed between Scylla - the danger of pre-arranged or tacit collusion if we use the Vickrey Rule - and Charybdis - inefficient allocation if we use another rule.

From a purely game theoretic point of view, the truth-revealing strategies are more convincing than collusion because truth-revealing is a dominant strategy. So let us assume, for the last remarks, that a Vickrey auction results in truth-revealing. Under this assumption, the Vickrey Auction is not only efficient but it is rather nice also under other aspects.

In all cases of inefficient auctions, there is the possibility of strategic bidding, particularly by large competitors. Under the Vickrey Rule, the large competitors have no advantage. So the American Treasury should have reacted to the Solomon Bros. scandal by introducing the Vickrey Rule instead of the single-price rule. In addition, under the Vickrey Rule, market entry is efficient. A new producer will enter the market only if he can beat the expected price of the marginal successful bid because that is the bid he will substitute and that is the price he will be paid.

In this paper, efficiency does not include transaction costs. We might find other rules with little loss of efficiency in the above sense and with low transaction costs. I believe, however, that the Vickrey auction could be optimal also under such a broader view - provided no collusion occurs.

Appendix

Proof of Proposition 1: Let us assume that (2) is not fulfilled. Then there are at least two bidders $i \neq k$ with different bidding functions b^i_j and b^k_l on a set with a positive measure. Take the sets

$S_1 = \{(x^i,x^k): x^i_j > x^k_l \text{ and } b^i_j(x^i) < b^k_l(x^k)\}$ and

$S_2 = \{(x^i,x^k): x^i_j < x^k_l \text{ and } b^i_j(x^i) > b^k_l(x^k)\}$ (see Fig. 1).

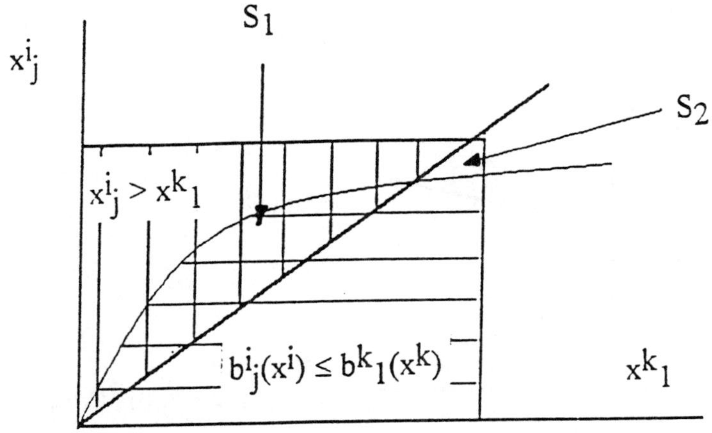

Figure 1: The order of values and the order of bids

At least one of these sets has a measure different from 0. Therefore - and because of the generally positive density function - the event "m-1 bids are smaller than b^i_j and b^k_1, all other bids are larger than b^i_j and b^k_1, and b^i_j and b^k_1 are either from S_1 or S_2" has a positive probability.

Thus, if (2) were not fulfilled there would be a positive probability of not selecting the m highest values.

A similar argument excludes that g is not strictly monotone.

Proof of Lemma 1: Let $C(b_{-j}, \beta)$ and $U(b_{-j}, \beta)$ denote the values of C and U when b_j is substituted by β. If g(.) is differentiable then, because of Assumptions 1 and 5, $P_j(.)$ and $C(b_{-j},.)$ are also differentiable. So, if U is non-differentiable then there is an isolated point of non-differentiability of all three functions. U can be discontinuous or it can be continuous but non-differentiable. We can argue similarly in both cases. So let us concentrate on the second case (see Figure 2).

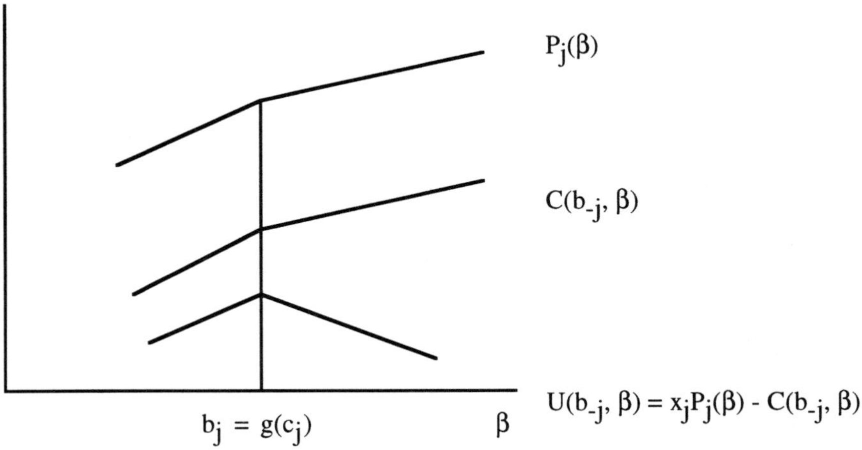

Figure 2: An (impossible) case of U being non-differentiable at $\beta = g(x_j)$.

Because $b_j = g(x_j)$ is a best reply to the bidding strategies of the others, U has to have a peak at b_j, i.e. the "derivative from the left side" is not smaller than 0 and larger than the "derivative from the right side" which is not larger than 0.

Let us assume that "the derivative from the left" is larger than 0. Then, by decreasing x_j by a small enough amount it remains larger than 0 while the derivative from the right is smaller then 0. Though x_j has changed, the maximum of U remains at the same place. That is a contradiction to $g(x_j)$ being an increasing function (Proposition 1).

The respective argument excludes that "the derivative from the right side" is smaller than 0. So both derivatives have to be 0, i.e. U is differentiable at $b_j = g(x_j)$.

Proof of Proposition 3: Let g(.) be the bidding rule connected with $c(\mathbf{b},\mathbf{w})$. Because of the Revenue Equivalence Theorem and because of

$$I(g^{-1}(w_{m-j+1}) < g^{-1}(b_j)) = I(w_{m-j+1} < b_j) \tag{A1}$$

it follows that

$$\begin{aligned} H(\mathbf{b}) &= C(b) - C^v(g^{-1}(b)) \\ &= \int [c(\mathbf{b},\mathbf{w}) - c^v(g^{-1}(b), g^{-1}(\mathbf{w}))]\, dF(\mathbf{w}) \\ &= \sum_{j=1}^{m} \int_{w_{m-j+1}=g(0)}^{b_j} \int_{w^{m-j+1}} [\ldots] dF(w) \\ &= 0 \end{aligned} \tag{A2}$$

with $\mathbf{w}^{m-j+1} = (w_1, w_2, \ldots, w_{m-j}, w_{m-j+2}, \ldots, w_m)$.

In particular (A2) implies

$$H(b_1,\ldots,b_{n-1},b_n)-H(b_1,\ldots,b_{n-1},g(0)) \tag{A3}$$
$$= \int_{w_1=g(0)}^{b_m} \int_{w^1} \left[c_m(b,w)-g^{-1}(w_1)\right] dF(w)$$
$$= 0.$$

Because c_m is monotone in b_k it must be independent of b_k, $k \neq m$, almost everywhere. Now we can apply the same reasoning to c_{m-1}. Because of (A2) and (A3) we get

$$H(b_1,\ldots,b_{n-1},b_n)-H(b_1,\ldots,b_{n-2},g(0),g(0)) \tag{A4}$$
$$= \int_{w_2=g(0)}^{b_{m-1}} \int_{w^2} \left[c_{m-1}(b,w)-g^{-1}(w_2)\right] dF(w)$$
$$= 0,$$

Thus c_{m-1} cannot depend on b_k, $k \neq m-1$. Formally the proof that c_j does not depend on b_k, $k \neq j$, can be closed by induction.

Let us now come to $c_j'(b_j,\mathbf{w})$. From

$$C_j(b_j) = \int_{w_{m-j}=g(0)}^{b_j} \int_{w^{m-j+1}} c_j(b_j,w)dF(w) \tag{A5}$$

and Proposition 2 follows

$$dC_j/db_j = C_j'(b_j) = \int_{w^{m-j+1}} c_j(b_j,w)dF(w/w_{m-j+1}=b_j) \tag{A6}$$
$$+ \int_{w_{m-j+1}=g(0)}^{b_j} \int_{w^{m-j+1}} c_j'(b_j,w)dF(w)$$
$$= g^{-1}(b_j)P_j'(b_j)$$

with

$$P_j(b_j) = \int_{w_{m-j+1}=g(0)}^{b_j} \int_{w^{m-j+1}} dF(w) \tag{A7}$$

$$dP_j/db_j = P_j'(b_j) = \int_{w^{m-j+1}} dF(w/w_{m-j+1}=b_j) \tag{A8}$$

$$P_j'(g(1)) = \Pr ob(w_k = g(1), k = 1, \ldots, m-j) \tag{A9}$$

$$= \begin{cases} 0 \; for \; j < m \\ 1 \; for \; j = m \end{cases}$$

Because of $c_j \geq 0$ and $c_j' \geq 0$, $P_j'(g(1)) = 0$ implies that both integrals of (A6) are equal to 0 for $b_j = g(1)$. Thus $c_j'(g(1),\mathbf{w}) = 0$ has to be fulfilled almost everywhere.

References

Bolle, F., (1993), No Revenue Equivalence in Multiple Bid Auctions. Discussion Paper, Frankfurt (Oder), 1993.

Bolle, F., (1992), Supply function equilibrium and the danger of Tacit Collusion. The case of the Spot Markets for Electricity: *Energy Economics*, April 1992, 94 - 102.

Bolle, F., (1991), Wettbewerb mit Angebotsfunktionen. *Zeitschrift für Wirtschafts- und Sozialwissenschaften, 111*, 553-576.

Branco, F., (1992), Optimal Multiple Unit Auctions of an Indivisible Good. Discussion Paper. Universidade Catolica Portuguesa.

Bulow, J., Roberts, J., (1989), The Simple Economics of Optimal Auctions. *Journal of Political Economy 97*, 1060-90.

Capel & Co, (1990), Reshaping the Electricity Supply Industry in England and Wales, London, Februar 1990.

Cox, J.C., Smith, V.L., Walker, J.M., (1984), Theory and Behavior of Multiple Unit Discriminative Auctions. *Journal of Finance XXXIX*, 983-1010.

Die Deutsche Bundesbank, (1989), Geldpolitische Aufgaben und Instrumente. *Sonderdruck der Deutschen Bundesbank Nr. 7.*

Dubey, P., Shubik, M., (1980), A Strategic Market Game with Price and Quantity Strategies. *Zeitschrift für Nationalökonomie 40*, 25 - 34

Engelbrecht-Wiggans, R., (1988), Revenue Equivalence in Multi-Object Auctions. *Economic Letters 26*, 15-19.

Fehr, N.-H.M. von der, Harbord, D., (1991), An Auction Approach to the Study of Spot Market Competition in a Deregulated Electricity Industry. Discussion paper, University of Oslo.

Green, R., (1991), Reshaping the CEGB Electricity Privatisation in the UK. *Utilities Policy 1*, 245-54.

Green, R., Newbery, D.M., (1991), Competition in the British Electricity Spot Market. Discussion Paper, Centre for Economic Policy Research, London.

Güth, W., (1986), Auctions, public tenders, and fair division games. *Mathematical Social Siences 11*, 283-294.

Güth, W., van Damme, E.E.C., (1986), A Comparison of Pricing Rules for Auctions and Fair Division Games. *Social Choice and Welfare 3*, 177-198.

Harris, M.H., Raviv, A., (1981), Allocation Mechanisms and the Design of Auctions. *Econometrica 49*, 1477-99.

Klemperer, P.D., Meyer, M.A., (1989), Supply Function Equilibria. *Econometrica 57*, 1243-1277.

Leininger, W., (1989), Escalation and Cooperation in Conflict Situations. *Journal of Conflict Resolution 33*, 231-254.

Maskin, E. and Riley, J.G. (1989). Optimal Multi-Unit Auctions. In: Hahn, F., *The Economics of Missing Markets*, Information, and Games. Clarendon Press, 312-335.

Myerson, R.B., (1981), Optimal Auction Design. *Mathematics of Operation Research 6*, 58-73.

Nautz, D. (1994) Bidding Behavior in the Zinstender of the German Bundesbank: Theory and Evidence. Manuscript, Freie Universität Berlin.

Newbery, D.M., (1991), Cacity-constrained Supply Function Equilibria: Competition and Entry in the Electricity Spot Market. Diskussionspapier, University of Cambridge.

Plum, M., (1992), Characterization and Computation of Nash-Equilibria for Auctions with Incomplete Information. *International Journal of Game Theory 20*, 393 - 418.

Robinson, M.S., (1985), Collusion and the Choice of Auction. *Rand Journal of Economics 16*, 141-145.

Rothkopf, M., Teisberg, T., Kahn, E., (1990),Why are Vickrey Auctions Rare? Journal of Political Economy 98, 94-109.

Scherer, F.M., (1980), Industrial Market Structure and Economic Performance. Rand McNally, Chicago.

Shubik, M., (1971), The Dollar Auction Game: A Paradox in Non-cooperative Behavior and Escalation. *Journal of Conflict Resolution 15*, 545-547.

Ungern-Sternberg, Th. von, (1988), Cartel Stability in Sealed Bid Second Price Auctions. *The Journal of Industrial Economics XXXVI*, 351-358.

Vickrey, W., (1961), Counterspeculation, Auctions, and Competitive Sealed Tenders. *Journal of Finance 16*, 8-37.

Weber, R.J., (1983), Multiple Object Auctions. In: Schotter, A., (Ed.), *Studies in Game Theory and Mathematical Economics*, New York University Press, New York, 165-191.

Wilson, R.B., (1979), Auction of Shares. *Quarterly Journal of Economics*, 675-689.

BALANCED EQUILIBRIUM[1]

A Model of Behaviour in Games Based on Continuity and Insufficient Reason

M. MARIOTTI
University of Manchester, School of Economic Studies, Manchester M13 9PL, UK

1. Introduction

In this paper I propose a model of decision making under strategic uncertainty (that is, uncertainty concerning what payoff-relevant decisions other agents are going to make). Although the model is axiomatic, I hope to convince the reader of its potential fruitfulness by exhibiting some specific examples in which it seems to perform well. It is worth specifying at the outset that I do not claim universal validity either for the model itself or for the axioms justifying it. The spirit in which it is proposed is one of 'methodological pluralism', in view of some shortcomings, both at the axiomatic level and at the application level, of the dominant Bayesian model.

To be more specific, in game theory the decision problem faced by the agents is generally viewed as a set of classical individual decision problems under uncertainty (where an agent in isolation is uncertain about the realization of one out of many 'states of nature'), to which some form of (mutual and common) knowledge structure has been added. This allows one to use the classic representation theorems of decision theory (notably, Savage's (1954) or Anscombe-Aumann's (1963))[2]. The uncertainty of a player can thus be expressed by means of subjective probabilitites, her preferences by means of a cardinal utility function and her behaviour as expected utility maximization. The specification of knowledge of rationality assumed will determine the domain of the players' uncertainty (see e.g.

---

[1] I wish to thank, with the usual disclaimers, Luca Anderlini, Ken Binmore. Yossi Greenberg, Robert Nau, Hamid Sabourian, Szolt Sandor and participants to a 'Quaker' seminar at Cambrige University and at the FUR VII conference in Olso for their comments. Special thanks go to Aagnus Hylland, my discussant in Oslo, for very penetrating comments and for correcting a mistake. Finally, I am grateful to all the inflexibly dogmatic critics of the ideas contained in this paper, for stimulating me to improve the exposition and to show the connection with 'orthodox' theory.

[2] For a very recent and authoritative reassertion of this 'Bayesian' view of game theory, see Hammond (forthcoming).

R. Nau et al. (eds.), Economic and Environmental Risk and Uncertainty, 145–161.

Aumann (1987), Bernheim (1984, 1985), Brandenburger and Dekel (1986, 1987), Nau and McCardle (1990), Pearce (1984), Tan and Werlang (1988)). In this sense it has been claimed that "the logical roots of game theory are in Bayesian decision theory" (Myerson, 1991, p. 5) or that the "theory of rational behaviour in game situations will be a direct generalization of the Bayesian theory of rational behaviour under uncertainty" (Harsanyi, 1977, p.11).

Recently, the Bayesian approach to game theory has come under attack for various reasons (e.g. Binmore (1992, 1993)). Elsewhere (Mariotti, 1995, 1996a)) I have tried at length to argue that this approach does not provide satisfactory logical roots to game theory. I find that some of the axioms necessary in order to express the players' uncertainty by means of probability distributions, in particular the ordering axiom, the dominance axiom and the 'sure thing principle', are somewhat suspect in a game theoretic context (quite independently of their status as decision theoretic axioms *per se*)[3].

But, even apart from this, it would seem of interest to study what equilibrium concepts can be derived when the players' uncertainty cannot be expressed by means of subjective probabilities in the presence of strategic uncertainty. Indeed, the axiomatic justifications of game theoretic equilibria in a Bayesian framework seem to lead to a rather unsatisfactory choice of alternatives: either strong assumptions on priors (such as common priors) are not made, and in this case rather 'coarse' solution concepts (like rationalizability) are obtained; or the dubious assumption of common priors or even common posteriors has to be made in order to obtain 'finer' solutions (like correlated and Nash equilibria)[4].

My starting point in this paper is the observation that, even though in the Bayesian approach to game theory both utilities and probabilities are, or should be[5], derived from the axioms, there is no logical difficulty in first describing a game in terms of utilities and only then tackle the problem of uncertainty. In other words, instead of describing a game in terms of basic consequences, one can derive the von-Neumann Morgenstern utilities of the agents by means of objective probabilitites and describe the game in terms of these utilities. Indeed, this is the standard description of a game. This introduces valuable information (about the players' intensity of preferences) in the game, which can be used in the successive step of imposing restrictions on the behaviour of the players. The crucial restriction I shall want to impose

---

[3]Hammond (forthcoming) has addressed, and in my view solved, some of the problems I posed in the cited papers.

[4]Rubinstein and Wolinsky (1994) have recently proposed their 'rationalizable equilibria' precisely with the aim of 'filling the gap' between Nash and Rationalizability. Mariotti (1996b) proposes another (this time Bayesian) model to address the same problem.

[5]See Borgers (1993).

is intended to incorporate in the game an arbitrarily small degree of uncertainty on the part of each player on the other players' payoffs.

This idea is by no means new. It has been used in some refinements of Nash equilibrium (e.g. with Fudenberg et al.'s (1988) notion of 'robustness' or in the Harsanyi and Selten (1988) theory). The novelty of the proposal here will merely consist in introducing the uncertainty, and the logic of the refinements, directly at the stage of the *definition* of an equilibrium, rather than confining it at the *refinement* stage.

In the next section I discuss the main feature of the model I am proposing. In section 3 the axioms and results are fully explained and discussed. Section 4 contains some applications; it is shown that in all cases the proposed solution concept yields the 'reasonable' results. Section 5 concludes and suggests a simple way to test the theory empirically against other theories. The proofs are contained in the Appendix.

2. Introducing Cardinality in the Equilibrium Concept

Within the Bayesian framework, Nash equilibrium, the leading solution concept in noncooperative game theory, is only justifiable on the basis of strong assumptions (e.g., Bernheim (1985), Brandenburger and Dekel (1987), Tan and Werlang (1988), Aumann and Brandenburger (1995)). We wish to argue that despite this, it is a rather weak solution concept relative to the amount of information contained in the cardinal payoff structure of a game. Indeed, it does not exploit the full strength of the utility representation theorems obtained by means of objective probabilities. Only by enlarging the pure strategy spaces to the set of randomizations on these sets is the cardinal utility representation fully made use of. Even though the utility functions of the players are cardinal, only ordinal comparisons are needed in order to define a Nash equilibrium in pure strategies[6], so that the set of pure strategy equilibria is invariant under ordinal transformations of the payoff functions. This is not very satisfactory, because it amounts essentially to throwing away potentially valuable information about the intensity of preferences.

Some authors, notably Harsanyi and Selten (1988), have exploited this degree of freedom in order to refine the equilibrium set. Their concept of risk dominance exploits in fact the cardinality of the utility functions. However, if risk-dominance arguments are to be taken seriously when it comes to refining the Nash equilibrium set, it seems that the same type of arguments should be taken equally seriously when the equilibrium is being defined. It is remarkable that even some Bayesian authors have noticed that, on the basis

---

[6]Osborne and Rubinstein (1994) is the first textbook to use this fact explicitly in the description of a game when only pure strategy Nash equilibria are used. In that case, the cardinal apparatus is skipped.

of cardinal considerations some strategies which are not played in any Nash equilibrium (or any 'good' Nash equilibrium) can appear quite reasonable. Consider the game in fig. 1 (numbers denote here utility units derived from a cardinal utility function).

FIGURE 1

	L	R
T	100,1	0,0
M	0,1	100,0
B	99,1	99,2

Strategy B is not part of any pure strategy Nash equilibrium. If we allow for mixed strategies, B is part of two equilibria which are both weak, in the Harsanyi sense[7], in that they involve Row randomizing exactly with probabilities (1/2,1/2) between either T and B or M and B. Moreover, in these mixed strategy equilibria the expected payoff of Row is 99, which coincides with her security level: the prescription that Row should randomize and 'run the risk' of getting zero, when she could get 99 for sure, seems therefore to be quite unconvincing (see e.g. Aumann (1985), who describes a similar game). So, for example, given that Column plays L, T gives an additional payoff unit with respect to B. But clearly there is no way to attach any meaning to such a difference, given the cardinal nature of the payoff measurement. Loosely and informally speaking, Row could have a very 'weak' preference for T (formally speaking it is impossible in this context to even define the notion of absolute intensity of preferences). But no matter how weak her preference is, the avoided loss deriving from playing B instead of T should Columnn play R is ninety-nine times the comparative gain in the alternative case. This can be claimed in a formal sense, since comparisons of differences are meaningful with a cardinal representation. The same gain/loss ratio holds when comparing strategies M and B. In the light of these considerations, B seems an eminently reasonable, and at least not an unreasonable, way of playing the game for Row given the uncertainty concerning the play of Column.

More generally, what seems unsatisfactory in the notion of a Nash equilibrium, once it is assumed that we can deal with cardinal utility functions, is that it does not take any account of comparisons between utility differences. As was noted before, these comparisons turn out to be relevant

[7] A Nash eqilibrium is said to be weak in this sense when some of the strategies that constitute it are not the only optimal responses to the other strategies. In this case the Nash equilibrium seems to be less compelling: why should the players who are indifferent between two or more strategies play exactly the strategies of the assumed Nash equilibrium?

when the mixed extension of the game is considered; but they are not 'built in' the definition of the Nash solution concept itself.

In my view, cardinal comparisons, if brought to the fore, should be used not simply to refine the Nash equilibrium set, but more fundamentally to define a notion of equilibrium. I will argue presently that the invariance of the solution concept with respect to positive affine transformations can, if added to certain other postulates, lead to a comparatively sharp equilibrium notion that avoids 'almost' completely the use of subjective probabilities.

3. Axioms and Solution Concept

In this section I put forth a set of (what are conventionally but somewhat pompously called) axioms that aim to avoid the difficulties alluded to before. In particular, I want to avoid (i) the assumption that a player has a preference ordering on her own strategy set[8], asking instead that the player should merely be able to make a choice; and (ii) the notoriously problematic 'sure-thing principle'. Or, more directly, I will avoid demanding that in every circumstance she is capable of transforming her uncertainty in a probability distribution on the strategy sets of her opponents. I will, however, assume that she should consider every alternative as equally likely in one, and only one, case: namely, the case in which all her opponents are indifferent between all their strategies. Moreover, I will assume that neither a linear transformation of the payoff functions of the game nor iterated elimination of (strongly) dominated strategies should affect the players' decisions. Finally, I will impose a continuity condition on the expectations and behaviour of the players.

First, some notation. I assume that a set of N-person games in strategic form, Γ, has been fixed, with typical element G. Such an element is composed of an N-ple of (nonempty compact subsets of some topological space) strategy sets $(S_i)_{i=1,...,N}$ and an N-ple of elements from the set $U(S_1,...,S_N)$ of all bounded continuous real-valued utility functions u_i: $S_1 \times S_2 \times ... \times S_N \rightarrow \Re$. From now on we write $G = (S_i,u_i)$ to denote a generic game in Γ. No 'outside' mixing is allowed: S_i is a fully comprehensive description of a player's feasible strategies. Moreover, let $\Sigma_i = \{S_i |$ there exists $G \in \Gamma$ with $G = (S_i,u_i)_{i=1,...,N}$ for some $u_i \in U(S_1,...,S_N)\}$. Given a game G $= (S_i,u_i) \in \Gamma$, a positive affine (linear) transformation of G is a game T(G) such that $T(G) = (S_i,T_i(u_i))$ for some set $\{T_i\}_{i=1,...,N}$ of positive affine (linear) transformations of the utility functions u_i. A strategy $s_i \in S_i$ is dominated if

---

[8]The problem with the ordering assumption is, in a nuthshell, this: since a player must choose a strategy, why should she bother to order the unchosen strategies? Or, as Wolfowitz (1962) puts it: "When a man marries he presumably chooses, from among all possible women, the one whom he likes best. Need he necessarily be able also to order the others in order of preference?"

there exists $s'_i \in S_i$ with $u_i(s'_i,s_{-i}) > u_i(s_i,s_{-i})$ for all $s_{-i} \in S_{-i}$, in standard notation. Given $G \in \Gamma$, the *reduced* game G^* is the game obtained from G by iterated elimination of dominated strategies. That is, define with recursion on t, for all i, $S_i(0) = S_i$ and $S_i(t+1) = \{s_i \in S_i(t) |$ there exist no $s'_i \in S_i(t)$ with $u_i(s'_i,s_{-i}) > u_i(s_i,s_{-i})$ for all $s_{-i} \in S_{-i}(t)\}$. Then $G^* = \cap_{t=1,\ldots,\infty}(S_i(t),u^*_i)$, where u^*_i is the restriction of u_i to the subset of undominated strategies. Finally, for all i, let $M(s_i,S_{-i})$ denote the expected payoff from playing s_i when the uniform distribution on the product strategy set S_{-i} is used. The axioms can now be introduced. A discussion will follow each group of axioms.

AXIOM 1 (Choice): for all i, there exists a nonempty correspondence $c_i: \Gamma \rightarrow \Sigma_i$, such that, for all $G = (S_i,u_i) \in \Gamma$, $c_i(G) \subseteq S_i$.

AXIOM 2 (Dominated Strategies): for all $G \in \Gamma$ and for all i, $c_i(G) = c_i(G^*)$.

AXIOM 3 (Linear Tranformations): for all i, for all $G \in \Gamma$ and positive linear transformation T(G), $c_i(G) = c_i(T(G))$.

These first three axioms are relatively uncontroversial.

Choice says that all players in any game can always select some strategy out of their strategy sets. Note that a player is not required to form an ordering on this set; nor does the choice of strategy need to be unique. This is a first difference from the Bayesian approach, according to which a player can produce a complete ranking of the strategies via expected utility maximization with respect to some subjective probability distribution. This would seem to be an excessively strong requirement, as to 'solve' a (reduced) game one does not need a complete ranking. The next two axioms begin to give a characterization of rational choice.

Dominated Strategies states the invariance of choice with respect to iterated elimination of (strongly) dominated strategies: in the absence of a probability distribution on the strategy sets of the opponents, this means that a player would never choose a dominated strategy, would never consider relevant for her decision an opponent's strategy which is dominated, would never play a strategy which is dominated when only the relevant strategies of the opponents are considered, and so on.

Linear Transformations postulates the invariance of choice with respect to this type of transformation. This axiom is quite weak, in that it is satisfied when, like in standard decision theory under uncertainty, cardinal representations are given, and *a fortiori* when the representation is unique only up to an order-preserving transformation. However, as a consequence of the following axioms the solution concept will satisfy the stronger notion of invariance with respect to positive affine transformations.

Before stating the next axiom I shall need to assume that the games are presented in a form such that all inessential strategies have been merged together. Two strategies s_i, s'_i are inessential if, for all j and all $s_{-i} \in S_{-i}$, $u_j(s_i,s_{-i}) = u_j(s'_i,s_{-i})$.

AXIOM 4 (Insufficient Reason): for all G, for all i, if $u_j(s) = u_j(s')$ for all $j \neq i$ and all $s,s' \in S$, then $c_i(G) = \{s_i \in S_i | s_i \in \text{argmax}(M(s'_i,S_{-i}))\}$.

Insufficient Reason is the only axiom that introduces (subjective) probababilities in the framework. It deals with the extreme case where all the opponents of some player are absolutely indifferent as to what strategy to play, since the payoff is the same for any possible strategy combination. This is surely a less uncontroversial axiom than the previous ones, but also one which has an established tradition in the theory of decision under uncertainty (where it goes usually under the name of 'symmetry': see e.g. Kelsey (1993)) and also in game theory. If there was any relevant reason for one player to attach more probability to the event that the opponents will play a specific strategy combination, this should necessarily be related to their payoffs, and should therefore be incorporated in them. In the game theory I am considering (whose validity is certainly not claimed to be universal) *all* motivations for a player's action are collapsed into her payoffs. In such a world it would be, it seems to me, a plain logical contradiction for player i to believe that player j is more willing to play a strategy rather than another, if the payoffs attached to these strategies are the same[9] (this is a main difference with the Savage 'states of nature': there is no 'motivation', therefore no payoff, to explain the realization of a particular state of nature. In that case, it is not a logical contradiction to consider one state more likely than another). Suppose that player i believed that j, who is indifferent between strategies A and B, will play A with probability 0.6 and B with probability 0.4. What reason could i offer for her subjective beliefs? Certainly not the reason that, in his opinion, j prefers this particular objective randomization. since he must be indifferent between all randomizations. In order to be willing to play any randomization. In general, any relevant reason for i's belief in favour of strategy A (for example, concerning j's education, personality and the like) would also be a reason for j's choice of strategy A, which would contradict the assumed indifference between A and B.

Harsanyi (1977) (and, in a more complex way, Harsanyi and Selten (1988)) makes this point in his postulate A5, where he introduces the concept of centroid strategy, and his explanation is a model of clarity and precision:

"The postulate asserts that [a player] will be equally likely to choose any one of ... 'equally good' strategies ... This postulate (or some similar postulate) follows from the customary operational definition of equality between two

[9]Note that this is not equivalent to claiming that an indifferent player will randomize uniformly.

utilities. For example, let s'_i and s''_i be two strategies ... If player i were found to choose (say) strategy s'_i more often, i.e. with a higher probability, than he chooses strategy s''_i, then this would be taken as an indication that he attaches a higher utility to s'_i than he attaches to s''_i. He would not be regarded as attaching the same utility to both strategies unless he were found to choose them with the same probability" (p. 114).

Some additional notation is needed before stating the last axiom. A sequence of games $\{G^n\}$ is a sequence $\{(S_i,u_i^n)\}$ (notice that the strategy spaces are fixed). I assume that some natural topology has been defined on the space of utility functions, so that convergence is well defined. Similarly, I assume that a natural topology has been defined on the strategy spaces S_i and that S is endowed with the product topology. In particular, if S_i is finite, it will be endowed with the discrete topology. Given a sequence $\{T_i^n\}$ of subsets of S_i, convergence to a set T_i means that: i) for any open set O containing T_i there exists n^* with $T_i^n \subseteq O$ for all $n > n^*$; and ii) for all $t \in T_i$ and open set O containing t there is n^* such that $T_i^n \cap O \neq \emptyset$ for all $n > n^*$.

AXIOM 5 (Continuity): let $\{G^n\}$ be a sequence of reduced games converging to G and such that $u_i^n(s) = u_i^m(s)$ for all $s \in S$ and all m,n. Define $c_i^n = c_i(G^n)$ for all n. Then the sequence $\{c_i^n\}$ converges to $c_i(G)$.

Continuity demands that a player should behave in a 'similar' way in any two games in which the payoffs of her opponents (not her own) are 'very close'[10]. Continuity could be interpreted as imposing a restriction both on the expectations and on the choices of a player. That is, games in which the opponents have very 'similar' utility functions should generate 'similar'

---

[10] A critic, although sympathetic with the upper-semicontinuity part of the axiom (part (i) of the convergence definition), expressed concerns with regard to the lower-semicontinuity aspect (part (ii) of the convergence definition). A clarification is therefore needed. First of all, notice that what is being asked is only continuity with respect to other players' payoffs, not one's own: in fact, the maps c_i satisying all the axioms will have to be *discontinuous* with respect to variations in one player's own payoff. Secondly, requiring only upper-semicontinuity is appropriate when players have a perfect knowledge of the opponents' payoffs and play best responses accordingly. Lower-semidiscontinuous behaviour emerges at the points of indifference between pure strategies: at such a point, slight perturbations cause a sudden 'implosion' of the best reply set. In a model of best replies against some probabilistic expectation of the opponents' optimal behaviour, when their payoffs are known precisely, it would therefore be appropriate to assume only upper-semicontinuity, because one's best reply set 'implodes' when those of the others do. But in the case considered here, where behaviour is not derived from a best reply structure, it would not make sense to limit oneself thus. If there is uncertainty about payoffs, a player can never be absolutely sure of when the opponents are exactly indifferent between some of their strategies, so that they cannot condition their *own* best reply on small variations of the opponents' payoffs. Therefore 'explosions' after small perturbations should be ruled out as well as 'implosions'. What, in fact, should be peculiar to 'implosions'? See however the remark after proposition 1 and corollary 2.

expectations; and 'similar' expectations should generate 'similar' choices. But one should bear in mind that in the above framework expectations have not been formally defined; choice depends directly on the observed payoff structure of the game.

The essential motivation for this axiom is that, if one admits even a vanishing degree of uncertainty of one player concerning the payoffs of the other players, any behavioural rule which is very sensitive to the exact specification of the opponents' payoffs should be abandoned. Indeed, what a player gets to know about the opponents in a cardinal context are simply von Neumann-Morgenstern utility ratios: in whatever way they are communicated to a player, they must be specified up to a finite number of decimals (say 2^{10}); hence the physical impossibility that they are known with absolute certainty.

This argument appears to be in agreement with a widespread view among game theorists. For example, it is similar in spirit to the Nash equilibrium refinements proposed by Fudenberg et al. (1988), which are also based on perturbations of the payoff functions. They observe:

"Our analyst is not prepared to reject an equilibrium that cannot be rejected in games that are nearby" (p.356).

In a similar vein, Harsanyi and Selten (1988) claim that

"Even if the players have as complete information about the payoff matrix of the game as they can possibly have, each player will always have some irreducible minimum of uncertainty about the other player's actual payoffs ... This is so because every person's utility function is subject to at least some, very small, unpredictable random fluctuations because of changes in his mood or perhaps a sudden urge to use one of his pure strategies in preference to his other pure strategy".(p.14-15).

In general, it seems fair to say that most game theorists are prepared to accept this sort of arguments when it comes to Nash equilibrium refinements. But one should recognize that if it is agreed that the opponents' payoffs are not known with complete certainty, this should affect the very definition of equilibrium, not merely serve to refine the equilibrium set. So, to use Fudenberg et al's (1988) words, if "the analyst is not prepared to reject equilibria that cannot be rejected in games which are nearby", why should she be prepared to *accept* equilibria that cannot be accepted in games which are nearby?

Why have the sequences of games been restricted to reduced games? Clearly, elimination of dominated strategies would not otherwise be consistent with continuity. Such a restriction on the sequences is a specification of the notion of closeness. I take the view that a game where dominated strategies are present cannot be strategically close to a reduced game. We must therefore avoid those perturbations which would transform a non dominated strategy into a dominated one.

PROPOSITION 1: let Γ be a class of games such that, for all $G = (S_i,u_i)$ in G, S_i is a Hausdorff space for all i. Then, for every i, there exists one and only one correspondence $c_i(.)$ consistent with axioms 1 through 5, defined by $c_i(G) = \{s_i \in S^{*}_{i} | \; s_i \in \text{argmax}(M(s'_i,S^{*}_{-i}))\}$ for all $G \in \Gamma$, where S^{*} denotes the strategy space of G^{*}.

COROLLARY 2: let Γ be the class of finite N-person games, with the discrete topology imposed on the strategy sets. Then, for all i, there exists one and only one correspondence $c_i(.)$ consistent with axioms 1 through 5, defined by $c_i(G) = \{s_i \in S^{*}_{i} | \; s_i \in \text{argmax}(M(s'_i,S^{*}_{-i}))\}$ for any $G \in \Gamma$, where S^{*} denotes the strategy space of G^{*}.

The equilibrium concept thus obtained will be called *balanced equilibrium*. More precisely, let a map $c_i(.)$ satisfying axioms 1 through 5 be called acceptable. Then, for a game G, the set of balanced equilibria is given by the Cartesian product of the $c_i(G)$, where c_i is the acceptable map for all i.

REMARK: a simple modification of the proof of proposition 1 yields the result that in generic finite games (i.e., in all games where there are no ties in the expected payoffs of each strategy given the uniform distribution on the opponents' strategies), the characterization only needs the upper-semicontinuity part of the continuity assumption. We remark this since, in discussions, we have noticed that assuming only upper-semicontinuity somehow meets with less hostility.

Existence conditions are quite standard.

PROPOSITION 3: for any finite game G, the balanced equilibrium set is nonempty.

For the existence of balanced equilibria in other spaces we need further standard conditions.

PROPOSITION 4: let $G = (S_i,u_i)$ be a game such that a) S_i is a compact topological space for all i and b) u_i is continuous in s_j for all j. Then the balanced equilibrium set is nonempty.

A straightforward but appealing characterization of the balanced equilibrium set is the following:

COROLLARY 5: let Γ be a class of finite games. Then, for all $G = (S_i,u_i) \in \Gamma$,
$c_i(G) = \{s_i \in S^{\cdot}_i | s_i \in \mathrm{argmax}(\Sigma u^{\cdot}_i(s'_i,s_{-i}))\}$,
where the summation is taken over the elements of $S^{\cdot}_i$.

In words, in a balanced equilibrium the players use those strategies which are maximal in the reduced game in the following ordering: a strategy A is better than another strategy B if and only if the sum of the payoff gains and losses obtained in each contingency by playing A rather than B is greater than zero. Or, in other words, if and only if the total utility (obtained by summing the payoffs obtained in every contingency) of strategy A is greater than the total utility of strategy B. We see then that even if cardinality has not been imposed explicitly with the axioms, it is a necessary consequence of them. In particular, cardinality has been imposed 'locally' in a specific class of games (with the Insufficient Reason) axiom, and the Continuity axiom 'spreads' it to all games.

Also, notice that, although we can use a particular ordering to *describe* the choice as the set of maximizers of this ordering, it is not being claimed that the players *use* this ordering to rank all strategies. I have characterized a choice function, and have said nothing about orderings on strategy sets.

4. Examples

To persuade the reader of the reasonableness of the balanced equilibrium concept, I will discuss here two cases in which it offers quite intuitive and convincing conclusions. In these particular cases, it refines the set of Nash equilibria in ways that are usually considered desirable.

The Nash demand game G^N
An amount x is to be divided among two (risk-neutral) players, who simultaneously and independently submit a bid, x_i, i = 1,2. If $x \geq x_1 + x_2$, then player i gets x_i, otherwise both get zero.

Here we have:

$$c_i(G^N) = \mathrm{argmax}\ x_i\ \mathrm{prob}(x - x_i \geq x_j) = \mathrm{argmax}\ x_i\ (x - x_i),$$

so that the unique pair of balanced equilibrium strategies is $x_1^* = x_2^* = x/2$. It so happens that although there is a continuum of Nash equilibria, applying the balanced equilibrium concept yields, uniquely, the 'natural' outcome. The reason is this. The balanced equilibrium is based on the idea that a complete uncertainty surrounds the other players' choices if they survive the iterated elimination process. This uncertainty is enough to make the 50-50 proposal 'the best bet', much in the same way as this was accomplished in

Nash's original paper by means of a stochastic frontier in the utility space; the main difference being that in Nash's case the uncertainty was introduced exogenously and not embodied in the solution concept itself.

Risk Dominance versus Payoff Dominance
Here I consider two-by-two coordination games with two Nash equilibria for which the strategy corresponding to the Pareto dominated one yields a higher 'coordination failure' payoff, as shown in figure 2 where $b > a > c > 0$ (notice that this framework includes 'stag-hunt' games such as described by Aumann (1989)).

FIGURE 2

	L	R
T	a,a	c,0
B	0,c	b,b

The Nash equilibrium (T,L) risk dominates (B,R) if and only if $a^2 > (b-c)^2$ (Harsanyi and Selten (1988)), or equivalently if and only if $a + c > b$. So we have the following

Fact: the balanced equilibrium is the Pareto dominant Nash equilibrium (B,R) unless it is risk-dominated by (T,L), in which case (T,L) becomes the balanced equilibrium.

So, unlike the Nash equilibrium concept, the balanced equilibrium concept always selects a unique outcome for this class of games. There is no need to invoke any 'outside' criterion of rationality like payoff or risk dominance in order to select either of the two outcomes. Moreover, whenever risk dominance and payoff dominance are in conflict, priority is given to the former criterion. This seems to be in accordance with common game theoretic reasoning, at least as reflected in Aumann (1989) and also Harsanyi and Selten (1988, p. 357-59), who claim that unless collective rationality is imposed directly, individual rationality will always drive towards risk-dominance.

5. Concluding Remarks

5.1 EMPIRICAL TESTING

The model I am proposing can be tested in a very simple way against the hypothesis of Nash behaviour on the part of the players. Consider the game of figure 3, with $1 > k > 0$.

FIGURE 3

	L	R
T	1+k,1	k,k
B	0,0	1,1+k

In this class of games the mixed strategy Nash Equilibrium prescribes to play T and R with probability (1+k)/2, and there are two pure strategy Nash equilibria, (T,L) and (B,R). The unique balanced equilibrium pair of strategies is (T,R). By setting $k = 2^{1/2}-1$ the Nash equilibrium probabilities of observing (T,R) is either 0, if the players play according to the pure strategy equilibria, or ½, if they play according to the mixed strategy equilibrium. On the contrary the expected frequency of (T,R) is 1 if the players play the balanced equilibrium.

In this way it is possible to compare the deviations of the empirical frequencies of (T,R) from the reference frequencies 1, ½ and 0 and obtain a clear test of real players' behaviour.

5.2 THE QUESTION OF 'CONSISTENCY'

The equilibrium concept I am proposing might be at first sight unappealing to many theorists, because it is seemingly 'inconsistent'. In fact, it may happen that the players choose strategies that are not best replies to any of the strategies that the opponents might play in equilibrium. The requirement of a theory's internal consistency is, for instance, used by Myerson (1991) in his informal arguments to justify Nash Equilibrium. But the criticism of inconsistency is incorrect. We started from the idea that the payoffs should be subjected to cardinal measurement and that each player lacks a completely accurate perception of the opponents' payoffs. But an arbitrarily small perturbation of a cardinal payoff function may alter the ordering of the strategies based on the calculation of the total utility yielded by each of them. This implies that no player can know with certainty the evaluation of the strategies given by any opponent, and therefore that no player can know with absolute precision the strategies recommended by the theory to the other players. All she knows with certainty, on the basis of our definition of continuity, are the dominance relations between strategies, and these are duly taken into account into her play. There can therefore be no question of inconsistency.

5.3 DOES BALANCED EQUILIBRIUM TAKE INTO ACCOUNT 'STRATEGIC BEHAVIOUR'?

This equilibrium concept goes, on the issue of interdependence, at least as far as other well-known concepts such as Iterated elimination of strongly dominated strategies (recall that this is one of the axioms in the present theory), Bernheim's (1984) and Pearce's (1984) Rationalizability, Basu and Weibull's (1991) notion of strategy subsets closed under rational behaviour, and Borgers' (1993) pure strategy dominance. It goes of course rather farther in the sharpness of the predictions it makes. Although I obviously accept the point that game theory must go beyond the theory of individual decision making[11], I also think it important not to underestimate the difficulties implicit in the attempt of going beyond some form or the other of iterated elimination of strategies. Once we try to do this, the assumptions needed are bound to be controversial, since they will not simply stem from knowledge of rationality and optimizing behaviour. The assumptions underlying Nash equilibrium, for example, are obviously inadequate in many situations, since they effectively require each player to know what the others are doing (e.g. Ellsberg (1956), Aumann and Brandenburger (1995)), yet this is the most used tool of game theory. The assumptions made here are no doubt also controversial, but I would like to make two final points in defence of the claim that they are worth considering.

First, they reflect an effort to arrive at a new equilibrium concept by incorporating notions which have already been used by game theorists when trying to refine the Nash equilibrium set (namely, continuity notions and Harsanyi's notion of centroid strategies). It would be strange if game theorists, who are willing to stomach such methodology in other circumstances, became suddenly squeamish with regard to the equilibrium notion which is being proposed.

Second, the results I offer in the examples speak, in a way, for themselves. Although the games I consider (Aumann's (1985) 'safe strategy game', the general coordination game and the Nash demand game) are very diversified in nature (one has a continuum of Pareto efficient pure strategy Nash equilibria, another has Pareto ranked pure strategy Nash equilibria, yet another has a unique pure strategy Nash equilibrium), the balanced equilibrium predicts exactly the type of behaviour which is thought of as 'reasonable' by the game theorists who singled out these games to illustrate some difficulties with the notion of Nash equilibrium.

References

Anscombe, F.J., and Aumann, R.J. (1963) A Definition of Subjective Probability, *Annals of Mathematical Statistics* **34**: 199-205.

Aumann, R. J. (1985) On the Non-Transferable Utility Value: a Comment on the Roth-Schafer examples, *Econometrica* **53**: 667-67.

Aumann, R.J. (1987) Correlated Equilibrium as an Expression of Bayesian Rationality, *Econometrica* **55**:1-18.

---

[11] And therefore disagree with views such as Kadane and Larkey's (1982).

Aumann, R. J. (1989) Nash Equilibria Are Not Self-Enforcing, Hebrew University of Jerusalem, mimeo.

Aumann, R.J. and A. Brandenburger (1995) Epistemic Conditions for Nash Equilibrium, *Econometrica* **63**: 1161-80.

Basu, K. and Weibull, J.W. (1991) Strategy Sets Closed under Rational Behaviour, *Economics Letters* **36**: 141-46.

Bell, D.E., Raiffa, H., and Tversky, A. (eds.) (1988) *Decision Making. Descriptive, Normative and Prescriptive Interactions*, Cambridge University Press, Cambridge.

Bernheim, D. (1984) Rationalizable Strategic Behaviour. *Econometrica* **52**:1007-28.

Bernheim, D. (1985) Axiomatic Characterizations of Rational Choice in Strategic Environments, *Scandinavian Journal of Economics* **88**:473-88.

Binmore, K. (1993) De-Bayesing Game Theory, in K. Binmore, A.P. Kirman and P. Tani (eds.) *Frontiers of Game Theory*, Cambridge, Ma, MIT Press.

Binmore K. (1992) Foundations of Game Theory, in J. J. Laffont (ed.), *Advances in Economic Theory*, Econometric Society Monographs, Cambridge University Press, Cambridge.

Borgers, T. (1993) Pure Strategy Dominance, *Econometrica* **61**: 423-30.

Brandenburger, A., and Dekel, E. (1986) On an Axiomatic Approach to Refinements of Nash Equilibrium, Economic Theory Discussion Papers, University of Cambridge, n^0. 104.

Brandenburger, A., and Dekel, E. (1987) Rationalizability and Correlated Equilibria, *Econometrica* **55**:1391-402.

Ellsberg, D. (1956) Theory of the Reluctant Duelists, *American Economic Review* **46**: 909-22.

Fudenberg, D., Kreps, D., and Levine, D. (1988) On the Robustness of Equilibrium Refinements, *Journal of Economic Theory* **44**:354-80.

Harsanyi, J. (1977) *Rational Players and Bargaining Equilibrium in Games and Social Situations*, CUP, Cambridge.

Harsanyi, J. and Selten, R. (1988) *A General Theory of Equilibrium Selection in Games*, The MIT Press, Cambridge, Ma.

Kadane, J. and P.D. Larkey (1982) Subjective Probability and the Theory of Games, *Management Science* **28**: 113-20.

Kelsey, D. (1993) Dutch Book Arguments and Learning In a Non-Expected Utility Framework, University of Birmingham, Dept. of Economics Discussion Paper 93-03.

Mariotti, M. (1995) Is Bayesian Rationality Compatible with Strategic Rationality?, *Economic Journal* **105**: 1099-1109.

Mariotti, M. (1996a) The Decision Theoretic Foundations of Game Theory, in K. Arrow, E. Colombatto, M. Pearlman and C. Schmidt (eds.) *The Rational Foundations of Economic Behaviour*, MacMillan, London.

Mariotti, M. (1996b) Qualitatively Consistent Strategic Behaviour, mimeo. University of Manchester.

Myerson, R.B. (1991) *Game Theory. Analysis of Conflict*, Harvard University Press, Harvard.

Nau, R.F. and K.F. McCardle (1990) Coherent Behaviour in Noncooperative Games, *Journal of Economic Theory* **50**: 424-44.

Pearce, D. (1984) Rationalizable Strategic Behaviour and the Problem of Perfection, *Econometrica* **52**:1029-50.

Savage, L.J. (1954) *The Foundations of Statistics*, John Wiley and Sons, New York.

Tan, J.C.C., and S.R.C. Werlang (1988) The Bayesian Foundations of Solution Concepts in Games, *Journal of Economic Theory* **45**:370-91.

von Neumann, J., and Morgenstern, O. (1947) *Theory of Games and Economic Behaviour*, Princeton University Press, Princeton, NJ.

Appendix: Proofs

PROOF OF PROPOSITION 1: existence is guaranteed by Choice. Fix a game $G = (S_i,u_i)$ and a player i. By Dominated strategies we can assume without loss of generality that G is a reduced game. For any natural number k, let (u_j/k) denote the utility function such that, for all $s \in S$, $(u_j/k)(s) = u_j(s)/k$. Consider the class of linear transformations $T^k(G)$ defined by $T^k(G) = (S_i,u_i,u_{-i}/k)$, where u_{-i}/k denotes the array of utility functions (u_j/k), $j \neq i$, with k a natural number. This class defines a sequence of games $\{G^k\}$ given by $G^k = T^k(G)$ for all k. Let 0_j be the utility function such that $0_j(s) = 0$ for all $s \in S$, and let $G^0(i) = (S_i,u_i,0_{-i})$ (in obvious notation). Then $\{G^k\}$ converges to $G^0(i)$. By Insufficient Reason $c_i(G^0(i)) = \{s^i \in S_i| \ s_i \in \text{argmax}(M(s'_i,S_{-i}))\}$. By Linear Transformations, the sequence $\{c_i^k\}$ given by $c_i^k = ci(G^k)$ for all k is constant and $c_i^k = c_i(G)$ for all k. Any constant sequence in a Haussdorf space converges uniquely to the constant value of the sequence. Therefore for all k $c_i^k = c_i(G^0(i))$ Then, Continuity implies that $c_i(G) = c_i(G^0(i))$. Q.E.D.

PROOF OF COROLLARY 2: since in this case we endow S_i with the discrete topology, S_i is Hausdorff. Q.E.D.

PROOF PROPOSITION 3: obvious

PROOF OF PROPOSITION 4[12]: without loss of generality, assume that S is the set which survived until the kth round of elimination. We show that if it is nonempty and compact, then the set of strategies surviving the next round is also nonempty and compact. This will show by induction, together with the

[12] I am grateful to Aagnus Hylland for reminding me that when asserting the openness of an intersection of open sets finiteness is of some importance.

compactenss and nonemptiness of the initial set, that at all rounds of elimination the strategy space is nonempty and compact.

The continuity of u_i in s_i for all i ensures that any open image of u_i has an open counterimage in S_i. Then, fixing $s_i \in S_i$ and $s_{-i} \in S_{-i}$, the set

$$D(s_i,s_{-i}) = \{s'_i \in S_i | \, u_i(s_i,s_{-i}) > u_i(s'_i,s_{-i})\}$$

is open. Define the set of strategies dominated by s_i,

$$D(s_i) = \bigcap_{s_{-i} \in S_{-i}} D(s_i, s_{-i}),$$

and the set of dominated strategies in S_i,

$$D_i = \bigcup_{s_i \in S_i} D(s_i),$$

If $D(s_i,s_{-i})$ is empty for all s_{-i}, then $D(s_i)$ is open. If $D(s_i,s_{-i})$ is nonempty for some s_{-i}, then it follows straightforwardly from the continuity of u_i in s_{-i} and from the compactness of S_{-i} that $D(s_i)$ can still be expressed as a *finite* intersection of sets $D(s_i,s_{-i})$, so that $D(s_i)$ is open for each s_i.

Therefore D_i, as a union of open sets, must be open as well. Consequently, at any stage of elimination the set of undominated strategies for i, $U_i = S_i \backslash D_i$ is closed. In addition, it is obviously nonempty (for each s_{-i}, the set of maximizers -hence undominated- of u_i on S_i is nonempty). As a nonempty closed subset of a compact set, U_i is then a nonempty compact set, and so is the Cartesian product over i of the U_i.

The intersection of a nested sequence of nonempty compact sets is nonempty. Moreover, the intersection of closed sets is closed, and therefore, as a subset of a compact set, compact. Thus, the reduced game $G^{\cdot}$ has a nonempty and compact strategy set $S^{\cdot}$. Continuity of u_i in s_i entails continuity of $M(s_i,S^{\cdot}_{-i})$, and compactness of $S^{\cdot}_{-i}$ yields the nonemptiness of the set of maximizers. Q.E.D.

PROOF OF COROLLARY 5: obvious.

MIXED EXTENSIONS OF GAMES AND THE SAINT PETERSBURG PARADOX

H.H. HALLER

Department of Economics

Virginia Polytechnic Institute and State University

Blacksburg, VA 24061, U.S.A.

The expected utility approach to the Saint Petersburg Paradox is extended to strategic games in normal form with a finite set of players.

1. Introduction

The so-called Saint Petersburg Paradox[1] concerns the fact that a person would pay only a modest amount of money for a lottery ticket L whose expected monetary payoff is infinite. One solution to that puzzle is to consider very small probabilities to be behaviorally insignificant, i.e. treat them as zeros. The prevailing solution is to replace expected monetary payoff maximization by expected utility maximization. That is, each monetary payoff x is assigned a (von Neumann-Morgenstern) utility or payoff $u(x)$; instead of the mathematical expectation $E_L x$, the mathematical expectation $E_L u(x)$ is used to evaluate the decision maker's subjective appreciation of the lottery. While a particular von Neumann-Morgenstern utility

[1] An extensive narrative of the history and the status of the paradox is given in Jorland (1987). See also Bassett (1987). An economic perspective is provided in Samuelson (1977). For a sample of recent additions, see the sequence Ryan (1974), Arrow (1974), Fishburn (1975) on the one hand, the sequence Shapley (1977a), Aumann (1977), Shapley (1977b) on the other hand, and Nielsen (1984, 1992) on the technical far side.

R. Nau et al. (eds.), Economic and Environmental Risk and Uncertainty, 163–171.

function u may make $E_L u(x)$ finite, there is always another lottery ticket T such that $E_T u(x)$ is undefined or infinite, unless u is bounded.

The very same issue — whether expected payoffs are always well defined and finite — re-emerges in game theory, with the traditional single person decision problem corresponding to the case of one player games. The formal setting is a strategic game in normal form with a finite set of players. THE QUESTION IS: Under what conditions does every joint mixed strategy give rise to a well defined and finite expected payoff to each of the players? The necessary and sufficient condition is that each player's payoff function on joint pure strategies be bounded. In other words, the boundedness condition is equivalent to the property that the set of all joint mixed strategies constitutes a — or rather is *the* maximal mixed extension of the game.

2. Example

To sketch the key argument, consider a lottery ticket L that pays 2^k with probability 2^{-k} for $k = 1, 2, \ldots$. Then $E_L x = \infty$. Now introduce the von Neumann-Morgenstern utility function $u(x) = \sqrt{x}$. Then $E_L u(x) = \sum_k 2^{-k} 2^{k/2} = \sqrt{1/2} \cdot (1 - \sqrt{1/2})^{-1}$. So the expected utility derived from L is finite. But since u is unbounded, another problem lurks around the corner. Take, for instance, the lottery ticket T that pays $s^k = 2^{4k}$ with probability 2^{-k}. Then $u(s^k) = 2^{2k}$ and $E_T u(x) = \infty$. Only boundedness of u would guarantee the absence of any further problem.

3. Model and Analysis

We first define a strategic game in normal form, the concepts of mixed strategy, admissible joint mixed strategy, mixed extension, and maximal mixed extension.

We use the conventional language and wisdom of integration theory.[2] After the basic definitions, we pose and analyze the fundamental problem.

In the sequel, $\Gamma = (I; (S_i)_{i\in I}; (u_i)_{i\in I})$ denotes a **strategic game in normal form**, that is:

1. I is a non-empty set of players;
2. S_i is a non-empty strategy set (strategy space), representing the pure strategies s_i for player $i \in I$;
3. $u_i : S \equiv \prod_{j\in I} S_j \longrightarrow \mathbb{R}$ is the payoff function for each player $i \in I$. S is called the space of joint pure strategies.

The following assumptions are maintained throughout.

(A) $I = \{1, \ldots, n\}$ for a fixed $n \in \mathbb{N}$.

(B) Each strategy space S_i is endowed with a σ-algebra $\mathcal{A}_i$.

Given (A) and (B), we write $S = S_1 \times \ldots \times S_n$ for the joint strategy space and $\mathcal{A} = \mathcal{A}_1 \otimes \ldots \otimes \mathcal{A}_n$ for the product σ-algebra on S. Let the real line $\mathbb{R}$ be endowed with its Borel σ-algebra $\mathcal{B}$. A further maintained assumption is

(C) Each u_i is $\mathcal{A}$-$\mathcal{B}$-measurable.

Definition. Let μ be a probability measure on $\mathcal{A}$ and $f : S \longrightarrow \mathbb{R}$ be $\mathcal{A}$-$\mathcal{B}$-measurable. Define $f^+ = \max\{0, f\}$ and $f^- = \max\{0, -f\}$. f is called **μ-integrable**, if both $\int f^+ d\mu$ and $\int f^- d\mu$ are finite. Then $\int f d\mu \equiv \int f^+ d\mu - \int f^- d\mu$.

Fact. f is μ-integrable if and only if $|f|$ is μ-integrable.

[2]For further details, consult Folland (1984).

Definition. For $i \in I$, a probability measure σ_i on $\mathcal{A}_i$ is called a **mixed strategy** for player i. Let Σ_i denote the space of mixed strategies for player i. We treat S_i as a subset of Σ_i, identifying $s_i \in S_i$ with the unit mass concentrated on s_i. Let $\Sigma = \Sigma_1 \times \ldots \times \Sigma_n$ denote the space of **joint mixed strategies**.

Definition. Identifying $\sigma = (\sigma_1, \ldots, \sigma_n) \in \Sigma$ with the product measure $\sigma_1 \otimes \ldots \otimes \sigma_n$ on $\mathcal{A}$, call the joint mixed strategy σ **admissible**, if for all $i \in I$, u_i is σ-integrable. Let $\Sigma^* \subset \Sigma$ denote the set of admissible joint mixed strategies. For $\sigma \in \Sigma^*$, $i \in I$, put $U_i(\sigma) = \int u_i(s) d\sigma(s)$.

Definition. A game $\Gamma^1 = (I; (\Sigma_i^1)_{i \in I}, (U_i^1)_{i \in I})$ is called a **mixed extension of** Γ, if

$$S_i \subset \Sigma_i^1 \subset \Sigma_i \text{ for all } i \in I;$$
$$\Sigma^1 = \Sigma_1^1 \times \ldots \times \Sigma_n^1 \subset \Sigma^*;$$
$$U_i^1(\sigma) = U_i(\sigma) \text{ for all } \sigma \in \Sigma^1,\ i \in I.$$

Definition. A game $\Gamma^1 = (I; (\Sigma_i^1)_{i \in I}, (U_i^1)_{i \in I})$ is called a **maximal mixed extension of** Γ, if Γ^1 is a mixed extension of Γ and there is **no** mixed extension $\Gamma^2 = (I; (\Sigma_i^2)_{i \in I}, (U_i^2)_{i \in I})$ of Γ with $\Sigma_i^1 \subset \Sigma_i^2$ for all $i \in I$ and $\Sigma_i^1 \neq \Sigma_i^2$ for some $i \in I$.

We pose the following QUESTION:

(1) When are all joint mixed strategies admissible, i.e. when does $\Sigma = \Sigma^*$ hold?

The identity $\Sigma = \Sigma^*$ amounts to avoiding Saint Petersburg Paradoxes in a strategic game setting. Question (1) is equivalent to:

(2) When is $\Gamma^* = (I; (\Sigma_i)_{i \in I}; (U_i)_{i \in I})$ **the** maximal mixed extension of Γ?

THE ANSWER is provided by the

Theorem. $\Sigma = \Sigma^*$ if and only if u_i is bounded for each $i \in I$.

PROOF. **"IF":** By definition, $\Sigma^* \subset \Sigma$. Suppose each u_i is bounded. Let $\sigma \in \Sigma$. Then each u_i is $\mathcal{A}$-$\mathcal{B}$-measurable and bounded, hence σ-integrable. Therefore, $\sigma \in \Sigma^*$. This yields $\Sigma \subset \Sigma^*$. Hence $\Sigma = \Sigma^*$.

"ONLY IF": Suppose some u_i is unbounded. Fix $j \in I$ with u_j unbounded. Without loss of generality, we may assume that u_j has no upper bound. (An analogous argument applies if u_j has no lower bound.) Since u_j does not have an upper bound, there exists a sequence $s^k = (s_1^k, \ldots, s_n^k)$, $k = 1, 2, \ldots$ in S such that $u_j(s^k) > 2^{2nk}$. For $i \in I$, $k \in \mathbb{N}$, let $\mu_{i,k}$ denote the measure that puts point mass 2^{-k} on s_i^k and set $\sigma_i \equiv \sum_k \mu_{i,k}$, a mixed strategy for i. Let $\sigma = (\sigma_1, \ldots, \sigma_n)$. Then for each $k \in \mathbb{N}$,

$$\begin{aligned} &\int u_j^+ d\sigma \geq \sigma(\{s \in S | u_j(s) \geq u_j(s^k)\}) \cdot u_j(s^k) \\ &\geq \mu_{1,k} \otimes \ldots \otimes \mu_{n,k}(\{s \in S | u_j(s) \geq u_j(s^k)\}) \cdot u_j(s^k) \\ &\geq (2^{-k})^n \cdot u_j(s^k) > 2^{-nk} \cdot 2^{2nk} = 2^{nk}. \end{aligned}$$

This shows $\int u_j^+ d\sigma = \infty$. Therefore, u_j is not σ-integrable. Thus $\sigma \in \Sigma$, $\sigma \notin \Sigma^*$, and consequently, $\Sigma \neq \Sigma^*$. Q.E.D.

4. Qualifications and Ramifications

In the previous section, the expected utility approach to the Saint Petersburg Paradox has been extended to n-person games. It is found that a bounded von Neumann-Morgenstern utility function is required, if unlimited randomization is deemed desirable and the players adhere to expected utility maximization. Of course, this finding is irrelevant if players or decision makers do not follow an expected utility criterion.

Game theory from its very beginning has adopted the expected utility paradigm — as documented in von Neumann and Morgenstern (1944). In the strategic game context, one often wishes to preserve the product structure of the joint strategy space. This is reflected by independent randomization and motivates the notion of a mixed extension. Alternative approaches could be appealing under certain circumstances. One could resort to generalized games (pseudo-games, abstract economies) where a player is allowed to randomize only to the extent that everybody's expected utility remains well defined and finite. One could consider correlated strategies. That case resembles the single decision maker scenario. In recent years, about a dozen attempts have been made to analyze strategic games without expected utility maximizing players, such as Crawford (1990); Dow and Werlang (1994); Eichberger and Kelsey (1994), a paper presented at this conference; Haller (1995).

In the traditional problem of a single player or decision maker, the counterpart of the Theorem has been established by Menger (1934). Moreover, in the traditional context, several suggestions have been made to circumvent the problem of unbounded utility. In the subsequent exposition I shall ignore more or less trivial measurability questions. Let us assume an *outcome space* S and denote by $\mathcal{P}$ the set of **probability distributions** on S and by $\mathcal{U}$ the set of **utility functions** defined on S. Then most of the suggestions alluded to amount to identifying a subset $\mathcal{E}$ of $\mathcal{U} \times \mathcal{P}$ such that

$$E_\sigma u = \int u d\sigma \in \mathbb{R} \text{ exists} \tag{1}$$

for all $(u, \sigma) \in \mathcal{E}$.

Special Cases obtain:

•• Fix u and put $\mathcal{E} = \{u\} \times \mathcal{Q}$ for some $\mathcal{Q} \subset \mathcal{P}$. For instance, $\mathcal{Q}$ = set of all 'simple' probability measures putting point mass on finitely many points. Or $\mathcal{Q}$ = set of all σ satisfying (1). A frequent case in financial economics combines u exhibiting constant absolute risk aversion and $\mathcal{Q}$ consisting of normal distributions. The present paper shows that the choice of $\mathcal{Q} = \Sigma$ requires boundedness of u.

- •• Fix a subset $\mathcal{Q}$ of $\mathcal{P}$. Specify *axioms* for a preference relation $\succ$ on $\mathcal{Q}$. Show existence of u with (1) for all $\sigma \in \mathcal{Q}$ and $\sigma \succeq \sigma' \Leftrightarrow E_\sigma u \geq E_{\sigma'} u$. Variants of this approach distinguish between actions and consequences.

- •• Impose specific conditions such that best alternatives or responses exist in each state of the world. See Kennan (1981).

But why bother? Why not allow $E_\sigma u \in \{-\infty, +\infty\}$? The answer is that this relaxation does not buy us much. If u lacks lower and upper bounds, then one can show along the lines of the "ONLY If" part of the Theorem, that there exists a σ with $\int u^+ d\sigma = +\infty$ and $\int u^- d\sigma = +\infty$ so that u is not σ-integrable even in an extended sense. If u lacks just one bound, then there exist σ and σ' such that σ (first order or otherwise) stochastically dominates σ' and $E_\sigma u = E_{\sigma'} u \in \{-\infty, +\infty\}$. Indeed, whenever a bounded expected utility representation is derived in the literature, one of the axioms stipulates that preferences reflect a version of stochastic dominance.[3]

5. References

Arrow, K.J. (1970) Exposition of the Theory of Choice Under Uncertainty, Ch. 2 in K.J. Arrow, *Essays in the Theory of Risk-Bearing,* North-Holland Publishing Company: Amsterdam and London.

Arrow, K.J. (1974) The Use of Bounded Utility Functions in Expected-Utility Maximization: Response, *Quarterly Journal of Economics* **88**, 136-138.

Aumann, R.J. (1977) The St. Petersburg Paradox: A Discussion of Some Recent Comments, *Journal of Economic Theory* **14**, 443-445.

[3]See the role of DOMINANCE for Arrow's (1970) UTILITY BOUNDEDNESS THEOREM and axioms (3:B:a), (3:B:b) in von Neumann and Morgenstern (1944).

Bassett, G.W., Jr. (1987) The St. Petersburg Paradox and Bounded Utility, *History of Political Economy* **19**, 517-523.

Crawford, V.P. (1990) Equilibrium Without Independence, *Journal of Economic Theory* **50**, 127-154.

Dow, J. and Werlang, S.R.C. (1994) Equilibrium under Knightian Uncertainty, *Journal of Economic Theory* **64**, 305-324.

Eichberger, J. and Kelsey, D. (1994) Non-Additive Beliefs and Game Theory, mimeo.

Fishburn, P.C. (1975) Unbounded Utility Functions in Expected Utility Theory, *Quarterly Journal of Economics* **90**, 183-188.

Folland, G.B. (1984) *Real Analysis: Modern Techniques and Their Applications*, John Wiley & Sons: New York et al.

Haller, H. (1995) Non-Additive Beliefs in Solvable Games, mimeo.

Jorland, G. (1987) The Saint Petersburg Paradox 1713-1937, Ch. 7 in L. Krüger, L.J. Daston, and M. Heidelberger (eds.), *The Probabilistic Revolution, Vol. 1*, The MIT Press: Cambridge, MA.

Kennan, J. (1981) The Existence of Expected Utility Maximizing Decisions When Utility is Unbounded, *Econometrica* **49**, 215-218.

Menger, K. (1934) Das Unsicherheitsmoment in der Wertlehre, *Zeitschrift für Nationalökonomie* **5**, 459-485.

Nielsen, L.T. (1984) Unbounded Expected Utility and Continuity, *Mathematical Social Sciences* **8**, 201-216.

Nielsen, L.T. (1992) The Utility of Infinite Menus, *Economics Letters* **39**, 43-47.

Ryan, T.M. (1974) The Use of Bounded Utility Functions in Expected-Utility Maximization: Comment, *Quarterly Journal of Economics* **88**, 133-35.

Samuelson, P.A. (1977) St. Petersburg Paradoxes: Defanged, Dissected, and Historically Described, *Journal of Economic Literature* **15**, 24-55.

Shapley, L.S. (1977a) The St. Petersburg Paradox: A Con Game? *Journal of Economic Theory* **14**, 439-442.

Shapley, L.S. (1977b) Lotteries and Menues: A Comment on Unbounded Lotteries, *Journal of Economic Theory* **14**, 446-453.

Von Neumann, J. and Morgenstern, O. (1944) *Theory of Games and Economic Behavior,* Princeton University Press: Princeton. Sixth paperback printing, 1990.

CONSUMPTION WITH FLUCTUATIONS IN PREFERENCE

KOJU NAKAZAWA* AND JOHN D. HEY†
*Seinan Gakuin University
Nishijin 6-chome, Sawara-ku, Fukuoka 814, Japan
† University of York
Heslington, York YO1 5DD, United Kingdom

This paper presents an infinite-horizon model of consumer behavior with flexible preference over durable goods to show that frequency of demand switching is not constant even if the flexibility of preference is constant. The main result of this model is that more frequent demand fluctuations are brought about by increase in wealth or income. Our model uses the stochastic dynamic programming method combined with expected utility theory. It is also shown, however, that qualitatively similar results can be obtained even when the assumption of extremely rational behavior is dropped.

1. Introduction

It is often said that changes in consumer confidence closely correlate to business fluctuations. On the other hand, changes in demand for durable goods are used as an indicator of business cycles. These facts strongly suggest that consumers' preferences are flexible. If consumers' preferences were fixed, there would no need for firms to spend much money on advertising or on periodic model changes. Nor would any informational externalities among consumers have significant meaning. Moreover, when preferences are flexible, consumption may fluctuate even if income and prices are constant, which eventually causes uncertainty and many economic problems. It is therefore useful to explore a model of consumer's behavior introducing fluctuating preference, especially related to decisions over durable good consumption.

In this paper we present a simple infinite-horizon model to investigate the relationship between flexibility in preference and demand fluctuation. The main result of our model is that frequency of demand fluctuation caused by flexible preference is not constant even if flexibility in

R. Nau et al. (eds.), Economic and Environmental Risk and Uncertainty, 173–192.

preference, expressed in a probability distribution function of a preference parameter, is constant.

We make our model as simple as possible. There are three consumption goods in our model: two of them are durable for two periods[1] and the other is perishable. Money also exists as a financial asset. In every period just one unit of one of the durable goods is consumed by each consumer according to her current preference. It is therefore necessary for each consumer to determine whether to switch from a durable good to another in every period.[2] This decision on demand switching depends on several factors, such as prices, wealth, income and how each consumer recognizes uncertainty in one's own preference.

We assume that consumer's horizon is infinite. And specification of a utility function to be introduced later makes it possible to assume consumers are expected utility maximizers. Then applying the familiar dynamic programming method, we investigate how this decision of switching responds to changes in exogenous variables. One of the important results is that frequency of switching increases with wealth or income. To put it in other words, wealth accumulation aggravates uncertainty in demand for durable goods and shortens the average length of durability of those goods. Reductions in durable good prices yield the same consequence. Demand for relatively inexpensive durable goods is therefore more volatile. These results together imply that the more rapidly an economy is growing, the more frequent demand fluctuations firms in the ecnomy are confronting.

Despite our specification the modeling of the process of demand fluctuation is barely tractable. The reason is that not only does the amount of wealth affect the frequency of switching. It also depends on consumer's positions, that is, whether a consumer owns one of the durable goods or not. If a consumer switches, it affects the consumption of the perishable good, hence the amount of end-of-period wealth, and alters her position. Since both effects are entangled with each other, the realizing consumption path is highly complicated. Further specification would be therefore required to apply our model to analyzing business cycles or other economic problems. In this sense our model is merely a first step toward a more realistic and tractable model.

Nevertheless, our model can grasp several important features of demand for durable goods determined by flexible preference because the results obtained are independent of any particular assumptions on causes of flexibility in preference. Our results are valid no matter what the cause of capricious behavior is.

In the next section we explain the framework of our model. Analytical results are derived in the third section. Then we show some results of a simulation on a specified case to clarify remaining ambiguous points in the analysis. The final section presents an alternative model to illustrate that our results does not necessarily depend on the assumption of rational behavior. Similar results can be derived as long as the consumer considers the tradeoff between current and future consumption including durable goods.

2. The Model

Our model deal with a decision problem in which there are two durable goods, x_1 and x_2, a perishable good, y, and money. x_1 and x_2 are durable for two periods. They do not suffer any physical quality depreciation within their lifetime, but there is no market for used goods. Every consumer consumes just one unit of one of the durables in each period. Their prices are p_1 and p_2 respectively. The price of y is assumed to be unity to avoid notational redundancy.

A representative consumer has some amount of money, w, as wealth at a beginning of period and receives a constant amount of income, m, in each period. The representative consumer has an infinite scope of decision making and her utility function is time-additive.

We specify consumer's utility function in a simple manner. Let α be a parameter such that $0 \leqq \alpha \leqq 1$. Then representative consumer's utility function is expressed in the form of $\alpha U(y)$ if she owns one unit of x_1 and consumes y, and it is $(1-\alpha)U(y)$ if she owns one unit of x_2. The utility index, α, is a stochastic variable changing independently from period to period, which the representative consumer anticipates with a subjective probability distribution function, $F(\alpha)$. We normalize $U(y)$ as $U(0) = 0$ and assume its concavity, i.e., $U'(y) > 0$ and $U''(y) < 0$.

We also assume that the consumer determines her consumption of y before α takes a certain value at the beginning of each period. After observing the current value of α, the consumer chooses a durable good according to it. This assumption is merely a matter of timing. The latter part of it, however, might be somewhat misleading. It does not mean that the consumer is ignorant of her own preference until finally making up her mind. On the contrary, the consumer always knows her preference but it is changeable for some reasons unspecified here, e.g., informatinal stimuli and social interaction. The situation in which a person recognizes variability of the person's own preference is nothing other than facing uncertainty. It is reasonable to apply expected utility theory to our decision problem under uncertainty, though it is not the only possible way.

We apply the familiar method of stochastic dynamic programing to analyze this decision problem.[3] The recursive functions in our model, however, are little complicated because of the existence of durable goods.

Two durable goods make three consumer's positions: having one unit of x_1, having one unit of x_2, and having no durable goods due to having consumed the same durable good during the last two consecutive periods. Clearly the consumer's decision depends on the current position. For instance, if a consumer has already had a unit of x_1, the problem is whether to hold it or switch to x_2. In this position it is possible for her to spend nothing on the durable good. However, if the consumer has neither, she has to purchase a new one. Therefore there exists a critical value of α in each position that determines which one to be consumed.

We denote the critical value in the case of having no durable good by α_0^*. Similarly, the critical values in the positions having x_1 and x_2 are α_1^* and α_2^* respectively. Hereafter the subscripts attached to variables, 0, 1 and 2, indicate consumer's position at the beginning of relevant period. Since the consumer determined the optimum amount of y before knowing the actual value of α, it is clear that expected utility is an increasing function of α when she chooses x_1, while her expected utility is a decreasing function of α if she chooses x_2. This fact means that there is only one switching point in each consumer's position.

Now we can express the decision problem by a triplet of recursive functions as follows.

$$V_0(w) = \max_{\alpha_0^*, y_0} \Big[\int_0^{\alpha_0^*} \{(1-\alpha)U(y_0) + \rho V_2(r(w+m-y_0-p_2))f(\alpha)\}d\alpha + \int_{\alpha_0^*}^1 \{\alpha U(y_0) + \rho V_1(r(w+m-y_0-p_1))f(\alpha)\}d\alpha \Big] \quad (1)$$

$$V_1(w) = \max_{\alpha_1^*, y_1} \Big[\int_0^{\alpha_1^*} \{(1-\alpha)U(y_1) + \rho V_2(r(w+m-y_1-p_2))f(\alpha)\}d\alpha + \int_{\alpha_1^*}^1 \{\alpha U(y_1) + \rho V_0(r(w+m-y_1))f(\alpha)\}d\alpha \Big] \quad (2)$$

$$V_2(w) = \max_{\alpha_2^*, y_2} \Big[\int_0^{\alpha_2^*} \{(1-\alpha)U(y_2) + \rho V_0(r(w+m-y_2))f(\alpha)\}d\alpha + \int_{\alpha_2^*}^1 \{\alpha U(y_2) + \rho V_1(r(w+m-y_2-p_1))f(\alpha)\}d\alpha \Big] \quad (3)$$

In this formulation $f(\alpha) = F'(\alpha)$, ρ is a subjective discounting factor with respect to time and r is one plus the interest rate. The argument of each V function is wealth at the beginning of corresponding period. We assume that r and ρ are constant, though we can derive similar results even dropping this assumption. Concavity of the V function is provable if $U(y)$ is concave. However, as the proof is extremely tedious, we are omitting it here. It should be also noted that the critical value in each position is also a choice variable because determination of it is equivalent to the choice between the two durable goods.

The former integral part in each recursive function is the expected utility gained by consuming a unit of x_2 and the latter is the expected utility in the case where x_1 is chosen. For example, the first part in the right hand side of (1) or (2) shows the expected utility the representative consumer gains when she buys a unit of x_2. It means switching in (2). In both cases the position in the following period is withone unit of x_2. In the meantime, the corresponding term in (3) indicates the expected utility when the consumer decides to keep a one-period old used x_2, i.e., not to spend on the durables. In this case, however, the consumer has to buy a unit of new durable good in the next period. Incidentally, fluctuating α in this model also generates flexibility in preference over all consumption.

Since the amount of y is chosen before given the value of α, one can rewrite the above recursive functions into a simpler form. To do so, we define two more functions, $G(\cdot)$ and $H(\cdot)$, as follows.

$$H(\alpha_0^*) \equiv \int_0^{\alpha_0^*}(1-\alpha)f(\alpha)d\alpha + \int_{\alpha_0^*}^{1}\alpha f(\alpha)d\alpha$$

$$= \bar{\alpha} + (1 - 2\alpha_0^*)F(\alpha_0^*) + 2G(\alpha_0^*) \qquad (4)$$

where $\bar{\alpha}$ is the stochastic mean of α and $G(\cdot)$ is defined as

$$G(\alpha_0^*) \equiv \int_0^{\alpha_0^*} F(\alpha)d\alpha$$

Clearly $G(\cdot)$ is an increasing function, $G(0) = 0$ and $G(1) = 1 - \bar{\alpha}$ for any distribution function. Since $H'(\alpha_0^*) = (1 - 2\alpha_0^*)f(\alpha_0^*)$, $H(\cdot)$ satisfies the following properties.

$$H(0) = \bar{\alpha} \qquad (5)$$

$$H(1) = 1 - \bar{\alpha} \qquad (6)$$

$$\bar{\alpha},\ 1 - \bar{\alpha} < H(1/2) = \bar{\alpha} + 2G(1/2) < 1 \qquad (7)$$

Using $H(\cdot)$ function, the recursive functions are rewritten as follows.

$$V_0(w) = \max_{\alpha_0^*, y_0} [H(\alpha_0^*)U(y_0) + \rho\{1-F(\alpha_0^*)\}V_1(r(w+m-y_0-p_1)) + \rho F(\alpha_0^*)V_2(r(w+m-y_0-p_2))] \qquad (1')$$

$$V_1(w) = \max_{\alpha_1^*, y_1} [H(\alpha_1^*) U(y_1) + \rho \{1-F(\alpha_1^*)\} V_0(r(w+m-y_1)) + \rho F(\alpha_1^*) V_2(r(w+m-y_1-p_2))] \quad (2')$$

$$V_2(w) = \max_{\alpha_2^*, y_2} [H(\alpha_2^*) U(y_2) + \rho \{1-F(\alpha_2^*)\} V_1(r(w+m-y_2-p_1)) + \rho F(\alpha_2^*) V_0(r(w+m-y_2))] \quad (3')$$

Then, we can derive equations which determine the critical values, α_0^*, α_1^* and α_2^*, respectively.

$$\alpha_0^* U(y_0) + \rho V_1(r(w+m-y_0-p_1)) = (1-\alpha_0^*) U(y_0) + \rho V_2(r(w+m-y_0-p_2)) \quad (8)$$

$$\alpha_1^* U(y_1) + \rho V_0(r(w+m-y_1)) = (1-\alpha_1^*) U(y_1) + \rho V_2(r(w+m-y_1-p_2)) \quad (9)$$

$$\alpha_2^* U(y_2) + \rho V_1(r(w+m-y_2-p_1)) = (1-\alpha_2^*) U(y_2) + \rho V_2(r(w+m-y_2)) \quad (10)$$

It is easy to interpret these equations. The left hand side of each equation is utility of x_1 when α happens to take the critical value, while the right hand side represents that of x_2. Since utility of x_1 is an increasing function of α while utility of x_2 is a decreasing function, these equations imply that demand switching is chosen when both utilities meet each other. In other words, the consumer switches when her utility reaches its minimum in each position.

Our argument need not to impose any specific tie-breaking rule on this model. Some proper tie-breaking rule, however, can be attached if necessary. For example, the consumer chooses x_1 with probability $1-F(\alpha_0^*)$ and x_2 with $F(\alpha_0^*)$ when the consumer has no durables and α takes the value of α_0^*, and so on.

It is straightforward to get the first order conditions for y_0, y_1 and y_2 as below.

$$H'(\alpha_0^*) U'(y_0) = \rho r [\{1-F(\alpha_0^*) V_1'(r(w+m-y_0-p_1)) + F(\alpha_0^*) V_2(r(w+m-y_0-p_2))] \quad (11)$$

$$H'(\alpha_1^*) U'(y_1) = \rho r [\{1-F(\alpha_1^*) V_0'(r(w+m-y_1)) + F(\alpha_1^*) V_2(r(w+m-y_1-p_2))] \quad (12)$$

$$H'(\alpha_2^*) U'(y_2) = \rho r [\{1-F(\alpha_2^*) V_1'(r(w+m-y_2-p_1)) + F(\alpha_2^*) V_0(r(w+m-y_2))] \quad (13)$$

The meaning of these conditions is the usual one concerning the marginal rate of substitution between current and future consumption.

Equations from (8) to (13) together give an optimal consumption path. This simultaneous equation system, however, seems somewhat peculiar because there is a dimensional gap between the first order conditions for three critical values and for the perishable good. The former is expressed in the dimension of level of utility while the latter is in the dimension of marginal utility. Of course, this gap stems from the assumption that only one unit of one of the durable goods is consumed. We will, however, show later that this dimensional gap does not cause any serious problem. Also we will examine the second order condition later.

3. Analytical Results of The Model

In this section we investigate mainly the effects of changes in exogenous variables on the choice variables, which will clarify how the frequency of demand switching is affected by changes in market circumstances. At first, however, we show the numerical order of α_0^*, α_1^*, and α_2^*.

When the representative consumer, for instance, has a unit of used x_1, expected utility in the future without switching in the current period must be larger than that in the case of switching, i.e.,

$$V_0(r(w+m-y_1)) > V_2(r(w+m-y_1-p_2)) \tag{14}$$

The reason is as follows. If, on the contrary,

$$V_0(r(w+m-y_1)) \leqq V_2(r(w+m-y_1-p_2)) \tag{15}$$

the value of a unit of used x_1 exceeds p_2. Because inequality (15) means that compensation to a person having no durables by the amount of p_2 is not enough for the person to become indifferent to the position having a unit of used x_2. This observation obviously contradicts rational behavior. Because a new durable good can be consumed for one more period, it is more valuable than a used good. Thus inequality (14) must hold. Then, (9) implies that $1-2\alpha_1^*$ is positive. Therefore,

$$0 \leqq \alpha_1^* < 1/2 \tag{16}$$

Similar arguments lead to the following proposition.

Proposition 1 $0 \leqq \alpha_1^* < \alpha_0^* < \alpha_2^* \leqq 1$ and $\alpha_1^* < 1/2 < \alpha_2^*$.

The equality in (16) applies if

$$U(y_1) \lesseqgtr V_0(r(w+m-y_1)) - V_2(r(w+m-y_1-p_2)) \tag{17}$$

When wealth is very small or the prices of durable goods are sufficiently high, the consumer will not switch from x_1 even if she gets no utility in the current period. α_1^* in this case is a corner solution and a discontinuity arises. This discontinuity, however, is innocuous to our analysis.

The above argument suggests that the probability of switching goes up when wealth accumulates. To put it in other words, larger wealth increases frequency of demand fluctuation caused by flexible preference. It is one of our main purposes to prove that this conjecture is true.

We prove it in the case where the representative consumer has a unit of x_1. Differentiating (9) and (12) with w, we get the next simultaneous equations.

$$(K_1+K_2)\frac{dy_1}{dw} + J\,f(\alpha_1^*)\frac{d\alpha_1^*}{dw} = K_2 \tag{18}$$

$$J\frac{dy_1}{dw} - 2U(y_1)\frac{d\alpha_1^*}{dw} = -\rho r\{V_2'(w_{+1}) - V_0'(w_{+1}-rp_2)\} \tag{19}$$

where

$$w_{+1} \equiv r(w+m-y_1)$$

$$J \equiv (1-2\alpha_1^*)U'(y_1) - \rho r\{V_2'(w_{+1}) - V_0'(w_{+1}-rp_2)\}$$

$$K_1 \equiv H(\alpha_1^*)U''(y_1) < 0$$

$$K_2 \equiv \rho r[\{1-F(\alpha_1^*)\}V_0''(w_{+1}-rp_2)+F(\alpha_1^*)V_2''(w_{+1})] < 0$$

To know the sign of dy_1/dw and $d\alpha_1^*/dw$ we have to clarify the sign of J. It should be noted that J corresponds to the cross effect between y_1 and α_1^*. Therefore to know the sign of J is also to affirm the second order condition of this decision problem. More concretely, if

$$2U(y_1)(K_1+K_2) + J^2 f(\alpha_1^*) < 0,$$

the optimality of y_1 and α_1^* is assured.

To check the sign of J we rewrite the recursive function again in a different form. Note that the equation (9), the first order condition for α_1^*, holds at the optimum point and what we want to know is the sign of J at the optimum point. For this purpose by substituting (9) into (2') we get a new objective function shown below that is basically equivalent to (2') or (2).

$$V_1(w) = \max_y \left[\{\bar{\alpha} + 2G(\alpha_1^*)\}U(y_1) + \rho V_0(r(w+m-y_1))\right] \tag{20}$$

where α_1^* should be treated as a function of y_1 defined by (9), that is, $V_1(w)$ is optimized solely by y_1. Thus we obtain the first order condition for y_1 from (20) which must yield the same result as (12).

$$\{\bar{\alpha}+2G(\alpha_1^*)\}U'(y_1) - \rho r V_0'(r(w+m-y_1)) + 2U(y_1)F(\alpha_1^*)\frac{\partial \alpha_1^*}{\partial y_1} = 0 \tag{21}$$

where $\partial \alpha_1^* / \partial y_1$ is given by (9) as

$$\frac{\partial \alpha_1^*}{\partial y_1} = \frac{J}{2U(y_1)} \tag{22}$$

Because y_1 is the only unknown in (21) it is possible to get dy_1/dw using this equation.

$$\{2U(y_1)(K_1+K_2) + f(\alpha_1^*)J^2 - 2U(y_1)F(\alpha_1^*)J\}\frac{dy_1}{dw} = 2U(y_1)K_2 \tag{23}$$

On the other hand, (18) and (19) yield another equation of dy_1/dw.

$$\begin{aligned}&\{2U(y_1)(K_1+K_2) + f(\alpha_1^*)J^2\}\frac{dy_1}{dw}\\ &\quad = 2U(y_1)K_2 - \rho r\{V_2'(w_{+1}-rp_2) - V_0'(w_{+1})\}f(\alpha_1^*)J\end{aligned} \tag{24}$$

These two equations, (23) and (24), must be equivalent to each other. It is easily confirmed that this equivalency requires $J = 0$ at the optimum point. The fact that J is null implies that

$$\rho r\{V_2'(w_{+1}-rp_2) - V_0'(w_{+1})\} = (1 - 2\alpha_1^*)U'(y_1) > 0 \tag{25}$$

It also implies that the second order condition is satisfied.

Now effects of increase in wealth on y_1 and α_1^* are clearly determined by (18) and (19).

$$0 < \frac{dy_1}{dw} = \frac{K_2}{K_1+K_2} < 1 \tag{26}$$

$$\frac{d\alpha_1^*}{dw} = \frac{(1-2\alpha_1^*)U'(y_1)}{2U(y_1)} \tag{27}$$

The second inequality in (26) means y is a normal good. As α_1^* is always less than a half, the result in (27) means that α_1^* gradually approaches to

1/2 as wealth increases. When α_1^* increases, frequency of switching from x_1 to x_2 also increases. This conclusion is valid even if $F(\alpha)$ does not coincide with the true distribution function because any distribution function is an increasing function of α.

Similarly the effects of wealth in other cases are obtained as listed below.

$$0 < \frac{dy_0}{dw}, \frac{dy_2}{dw} < 1 \tag{28}$$

$$\frac{d\alpha_2^*}{dw} = \frac{(1 - 2\alpha_2^*)U'(y_2)}{2U(y_2)} > 0 \tag{29}$$

$$\frac{d\alpha_0^*}{dw} = \frac{(1 - 2\alpha_0^*)U'(y_0)}{2U(y_0)} \gtreqless 0 \Leftrightarrow \alpha_0^* \lesseqgtr \frac{1}{2} \tag{30}$$

Thus we get the second proposition.

Proposition 2 When wealth increases, (a) the amount of consumed perishable good increases in all consumer's positions, (b) frequency of switching from a durable good to another also increases in all consumer's positions, and (c) demand for each durable good without having a used one is more randomized even if the prices of the durable goods are different and $f(\alpha)$ is not symmetrical with respect to 1/2

It should be noted here again that the result that all critical values of α asymptotically converge to one-half is valid no matter what the subjective distribution function is. In other words, any difference between the two durable goods loses its meaning when wealth becomes larger and larger.

Imagine that every consumer's wealth in an economy simultaneously rises. The average length of durability will be shortened by more rapid switching. Frequent switching eventually worsens wasted resources unless there are complete markets for used goods. Consumers, however, benefit from the opportunity of switching compared to the case where just one durable good exists. On the other hand, markets for used goods are provided only imperfectly in the real economy due to uncertainty and transaction costs. One may conclude that enhancing consmers' welfare is not always compatible with effective resource allocation in the real economy.

If $F(\alpha)$ positively correlates with α in the previous period, frequency of demand switching will be smaller. Nevertheless, any correlation will not reverse the results derived above.

The prices of durable goods also affect consumer's decision making.

Effects of p_2 on y_1 and α_1^* are directly derived using (9) and (12).

$$\frac{dy_1}{dp_2} = \frac{-\rho\, rV_2''(w_{+1}-rp_2)}{K_1+K_2} < 0 \tag{31}$$

$$\frac{d\alpha_1^*}{dp_2} = \frac{-\rho\, rV_2''(w_{+1}-rp_2)}{2U(y_1)} < 0 \tag{32}$$

On the other hand, effects of changes in p_1 can not be derived so easily because the objective function includes p_1 only implicitly. We can, however, modify $V_0(w_{+1})$ and $V_2(w_{+1}-rp_2)$ as follows by applying the same procedure to derive (20).

$$V_0(w_{+1}) = \{\bar{\alpha}+2G(\alpha_0^*)\}U(y_0) + \rho\, V_1(r(w_{+1}+m-y_0-p_1))$$

$$V_2(w_{+1}-rp_2) = \{\bar{\alpha}+2G(\alpha_2^*)\}U(y_2) + \rho\, V_1(r(w_{+1}-rp_2+m-y_0-p_1))$$

Applying the first order conditions, we get the following:

$$\frac{\partial V_0(w_{+1})}{\partial p_1} = -\rho\, rV_1'(r(w_{+1}+m-y_0-p_1)) < 0$$

$$\frac{\partial V_2(w_{+1}-rp_2)}{\partial p_1} = -\rho\, rV_1'(r(w_{+1}-rp_2+m-y_0-p_1)) < 0$$

which determine the effect of changes in p_1 on y_1 as below.

$$\frac{dy_1}{dp_1} = \frac{-\rho\, r}{K_1+K_2}[\{1-F(\alpha_1^*)\}\frac{\partial V_0(w_{+1})}{\partial p_1} + F(\alpha_1^*)\frac{\partial V_2(w_{+1}-rp_2)}{\partial p_1}] < 0 \tag{33}$$

Furthermore, equation (25) asserts that

$$\frac{\partial V_2(w_{+1}-rp_2)}{\partial p_1} < \frac{\partial V_0(w_{+1})}{\partial p_1} < 0$$

which determines the effect of changes in p_1 on α_1^*.

$$\frac{d\alpha_1^*}{dp_1} = \frac{\rho}{2U(y_1)}\{\frac{\partial V_2(w_{+1}-rp_2)}{\partial p_1} - \frac{\partial V_0(w_{+1})}{\partial p_1}\} < 0 \tag{34}$$

Similar procedure gives the results summarized below.

$$\frac{dy_0}{dp_1},\ \frac{dy_0}{dp_2},\ \frac{dy_2}{dp_1},\ \frac{dy_2}{dp_2} < 0 \tag{35}$$

$$\frac{d\alpha_0^*}{dp_1} > 0, \quad \frac{d\alpha_0^*}{dp_2} < 0 \tag{36}$$

$$\frac{d\alpha_2^*}{dp_1} > 0, \quad \frac{d\alpha_2^*}{dp_2} > 0 \tag{37}$$

These analytical results together lead to the third proposition.

Proposition 3 If p_1 (p_2) is raised, it reduces new demand for x_1 (x_2) and it also reduces the frequency of switching from x_1 to x_2 (from x_2 to x_1). And any increase in the prices reduces the amount of perishable good consumption.[4]

This proposition means that simultaneous rises in p_1 and p_2 depress demand for durable goods as a result of negative wealth effect. Meanwhile, price reduction has positive wealth effect. Then price reduction may boost demand for durable goods, though volatility in demand fluctuation can be also worsened.

Price competition, however, has an important meaning in our model. If a monopolistic firm supplies those durable goods under our assumptions, i.e., one-unit consumption of the durable goods with the specified utility function, the firm will be able to raise the prices until exhausting almost all the consumer's purchasing power. We eliminate such a pathological case by implicitly assuming existence of price competition, i.e., there are at least two independent firms.

4. A Simulation

The argument so far suggests that y_0 is less than y_1 and y_2, i.e., the perishable good consumed in the position 0, having no durables, is less than those in the other positions, if other things are equal. It is provable that $y_0 < y_1$ by using the expressional form similar to (20).

$$V_0(w) = \{\bar{\alpha}+2G(\alpha_0^*)\}U(y_0) + \rho V_1(r(w+m-y_0-p_1))$$

$$V_2(w) = \{\bar{\alpha}+2G(\alpha_2^*)\}U(y_2) + \rho V_1(r(w+m-y_2-p_1))$$

Since $G(\alpha_0^*)$ is less than $G(\alpha_2^*)$, y_2 must be greater than y_0. Similarly, we can rewrite $V_0(w)$ and $V_1(w)$ in another form.

$$V_0(w) = \{\bar{\alpha}+(1-2\alpha_0^*)+2G(\alpha_0^*)\}U(y_0) + \rho V_2(r(w+m-y_0-p_2))$$

$$V_1(w) = \{\bar{\alpha}+(1-2\alpha_1^*)+2G(\alpha_1^*)\}U(y_1) + \rho V_2(r(w+m-y_0-p_2))$$

Since $\alpha_1^* > \alpha_0^*$, $(1-2\alpha_0^*)+2G(\alpha_0^*) < (1-2\alpha_1^*)+2G(\alpha_1^*)$. Then y_0 is smaller than y_1. It is, however, less obvious whether the difference among y_0, y_1 and y_2 will diminish or not when wealth increases, which closely relates to another ambiguous point in our model.

It is clear that $V_0(w)$ is less than $V_1(w)$ and $V_2(w)$. It is also clear that $V_1(w)$ is greater than $V_2(w)$ if p_1 is higher than p_2, and vice versa, because having a more expensive good is more advantageous. Meanwhile, the fact that all critical values converge to one-half implies that value of used goods are diminishing when wealth increases. If so, differences among $V_0(w)$, $V_1(w)$ and $V_2(w)$ will also be decreasing. These conjectures, however, are not necessarily guaranteed. The optimum conditions do not necessarily require the diminishing value of used goods. One could argue that value of used goods might be increasing with wealth due to the complimentarity between the durable good and the perishable good.

Simulation analysis is a useful tool to check these points. We show some results of a simulation here (Tables 1 to 4). We assumed a constant degree of absolute risk aversion, 0.05, in this simulation. Of course, any positive constant provides qualitatively similar consequences. We also assumed that $F(\alpha)$ is a uniform distribution function. As to the other parameters, $m = 6.0$, $\rho = 0.9$, and $r = 1.05$. Table 1 shows results in the case where p_1 and p_2 are 2.5 while in Table 4 they are 2.25. In Table 2 and 3 one of the prices is 2.25 and the other is 2.5.

The figures in these tables support our conjecture stated above as well as the results derived in the previous section. In all cases y_0 is less than y_1 and y_2. The differences among y_0, y_1 and y_2 are diminishing and the gaps among three V functions are also shrinking when w increases. α_0^*, α_1^* and α_2^* asymptotically converge to 0.5 in all tables. The effects of changes in prices are also observed. For instance, the effects of changes in p_2 is easily seen by comparing Table 2 to Table 4. When p_2 goes up from 2.25 to 2.5, all V functions shift downward but $V_2(w)$ remains highest. The perishable good consumption decreases in all consumer's positions, and $\alpha_0^*$and α_1^* fall while α_2^* rises.

The figures in these tables also give information about price elasticity of demand for the durable goods. In spite of our specification that just one unit of the durable good is consumed, switching behavior makes the elasticity positive. Then it is possible for firms to reduce the prices as a strategy in competition. As mentioned before, price competition is essential in our model. There is, however, an obstacle which may deter firms from reducing the prices. If all firms cut down their prices simultaneously, it does not affect α_0^* at all because the effects mutually cancel out, hence no immediate increase in demand for each durable good. α_0^* is always 0.5 in so far as both prices are equal and $f(\alpha)$ is symmetrical with respect to 1/2, as is seen in Table 1 and 4. Sufficient increase in

TABLE 1 ($\rho = 0.9$, $r = 1.05$, $m = 6.0$, $p_1 = 2.5$, $p_2 = 2.5$)

wealth	V_0	V_1	V_2
4.0	36.312645	36.858063	36.858063
8.0	38.715071	39.209011	39.209011
12.0	40.936702	41.405326	41.405326
16.0	43.045757	43.492577	43.492577
20.0	45.056728	45.481201	45.481201
24.0	46.984184	47.391613	47.391613
28.0	48.836372	49.229118	49.229118
32.0	50.614546	51.002163	51.002163
36.0	52.353027	52.718536	52.718536
40.0	54.026703	54.380829	54.380829
44.0	55.650959	55.994114	55.994114
48.0	57.227474	57.561100	57.561100
52.0	58.758812	59.083405	59.083405
56.0	60.246193	60.562058	60.562058
60.0	61.689384	61.996891	61.996891
	α_0^*	α_1^*	α_2^*
4.0	0.500000	0.421940	0.578060
8.0	0.500000	0.435762	0.564238
12.0	0.500000	0.444885	0.555115
16.0	0.500000	0.450891	0.549109
20.0	0.500000	0.455889	0.544111
24.0	0.500000	0.459452	0.540548
28.0	0.500000	0.462473	0.537528
32.0	0.500000	0.465034	0.534966
36.0	0.500000	0.467029	0.532971
40.0	0.500000	0.468729	0.531271
44.0	0.500000	0.470205	0.529795
48.0	0.500000	0.471438	0.528562
52.0	0.500000	0.472508	0.527492
56.0	0.500000	0.473466	0.526533
60.0	0.500000	0.474341	0.525659
	y_0	y_1	y_2
4.0	6.137341	6.479613	6.479613
8.0	7.385149	7.573560	7.573560
12.0	8.464518	8.626294	8.626294
16.0	9.395159	9.481400	9.481400
20.0	10.219389	10.325996	10.325996
24.0	10.963477	11.000000	11.000000
28.0	11.639156	11.746174	11.746174
32.0	12.241733	12.344685	12.344685
36.0	12.764727	12.867107	12.867107
40.0	13.205732	13.312169	13.312169
44.0	13.567922	13.678748	13.678748
48.0	13.851706	13.964184	13.964184
52.0	14.051597	14.169356	14.169356
56.0	14.175179	14.299006	14.299006
60.0	14.222566	14.356081	14.356081

TABLE 2 ($\rho = 0.9$, $r = 1.05$, $m = 6.0$, $p_1 = 2.25$, $p_2 = 2.5$)

wealth	V_0	V_1	V_2
4.0	36.898903	37.387436	37.442051
8.0	39.279819	39.730854	39.771935
12.0	41.495342	41.914291	41.957645
16.0	43.589535	43.990741	44.031147
20.0	45.590183	45.971195	46.009556
24.0	47.506027	47.872009	47.909477
28.0	49.348156	49.700329	49.736980
32.0	51.126461	51.465343	51.500687
36.0	52.845482	53.173447	53.207615
40.0	54.510376	54.827744	54.861073
44.0	56.125824	56.433430	56.465954
48.0	57.693966	57.993176	58.024609
52.0	59.217464	59.508648	59.539112
56.0	60.697521	60.980831	61.010296
60.0	62.133880	62.409706	62.438030
	α_0^*	α_1^*	α_2^*
4.0	0.489422	0.424922	0.568639
8.0	0.493957	0.436315	0.557004
12.0	0.494002	0.445523	0.548484
16.0	0.494877	0.451250	0.543435
20.0	0.495449	0.456248	0.539150
24.0	0.495763	0.459510	0.536056
28.0	0.496087	0.462742	0.533331
32.0	0.496376	0.465273	0.531096
36.0	0.496610	0.467264	0.529346
40.0	0.496784	0.468974	0.527816
44.0	0.496926	0.470423	0.526514
48.0	0.497073	0.471650	0.525441
52.0	0.497187	0.472706	0.524502
56.0	0.497282	0.473616	0.523684
60.0	0.497367	0.474517	0.522881
	y_0	y_1	y_2
4.0	6.197487	6.679082	6.689885
8.0	7.469510	7.621792	7.789722
12.0	8.603884	8.727019	8.748011
16.0	9.487163	9.565713	9.606927
20.0	10.303335	10.399332	10.433812
24.0	11.030472	11.100551	11.142565
28.0	11.739659	11.829917	11.848709
32.0	12.342417	12.432027	12.446016
36.0	12.867683	12.958650	12.967947
40.0	13.312837	13.407552	13.411134
44.0	13.679156	13.776645	13.777313
48.0	13.970053	14.061855	14.068399
52.0	13.176296	14.264628	14.276290
56.0	14.301455	14.393785	14.412094
60.0	14.354149	14.447618	14.471824

TABLE 3 ($\rho = 0.9$, $r = 1.05$, $m = 6.0$, $p_1 = 2.5$, $p_2 = 2.25$)

wealth	V_0	V_1	V_2
4.0	36.898903	37.442051	37.387436
8.0	39.279819	39.771935	39.730854
12.0	41.495342	41.957645	41.914291
16.0	43.589535	44.031147	43.990741
20.0	45.590183	46.009556	45.971195
24.0	47.506027	47.909477	47.872009
28.0	49.348156	49.736980	49.700329
32.0	51.126461	51.500687	51.465343
36.0	52.845482	53.207615	53.173447
40.0	54.510376	54.861073	54.827744
44.0	56.125824	56.465954	56.433430
48.0	57.693966	58.024609	57.993176
52.0	59.217464	59.539112	59.508648
56.0	60.697521	61.010296	60.980831
60.0	62.133880	62.438030	62.409706

	α_0^*	α_1^*	α_2^*
4.0	0.510578	0.431361	0.576078
8.0	0.506343	0.442996	0.563685
12.0	0.505998	0.451516	0.554477
16.0	0.505123	0.456467	0.548750
20.0	0.504551	0.460850	0.543752
24.0	0.504237	0.463945	0.540489
32.0	0.503624	0.466669	0.537258
36.0	0.503390	0.468905	0.534726
40.0	0.503216	0.470654	0.533700
44.0	0.503073	0.473485	0.529557
48.0	0.502927	0.474560	0.528350
52.0	0.502812	0.475498	0.527294
56.0	0.502718	0.476351	0.526340
60.0	0.502633	0.477119	0.525483

	y_0	y_1	y_2
4.0	6.197487	6.689885	6.679082
8.0	7.469510	7.789722	7.621792
12.0	8.603884	8.748011	8.727019
16.0	9.487163	9.606927	9.565713
20.0	10.303335	10.433812	10.399332
24.0	11.030472	11.142565	11.100551
28.0	11.739659	11.848709	11.829917
32.0	12.342417	12.446016	12.432027
36.0	12.867683	12.967947	12.958650
40.0	13.312837	13.411134	13.407552
44.0	13.679156	13.777313	13.776645
48.0	13.970053	14.068399	14.061855
52.0	13.176296	14.276290	14.264628
56.0	14.301455	14.412094	14.393785
60.0	14.354149	14.471824	14.447618

TABLE 4 ($\rho = 0.9$, $r = 1.05$, $m = 6.0$, $p_1 = 2.25$, $p_2 = 2.25$)

wealth	V_0	V_1	V_2
4.0	37.554245	38.033070	38.033070
8.0	39.908371	40.348854	40.348854
12.0	42.102970	42.519680	42.519680
16.0	44.164742	44.578346	44.578346
20.0	46.167545	46.544800	46.544800
24.0	48.070969	48.432148	48.432148
28.0	49.900265	50.248748	50.248748
32.0	51.666615	52.002041	52.002041
36.0	53.374474	53.698681	53.698681
40.0	55.028923	55.343098	55.343098
44.0	56.633949	56.938953	56.938953
48.0	58.192810	58.489101	58.489101
52.0	59.707161	59.995621	59.995621
56.0	61.178302	61.458992	61.458992
60.0	62.606380	62.879772	62.879772
	α_0^*	α_1^*	α_2^*
4.0	0.500000	0.432058	0.567961
8.0	0.500000	0.444433	0.555567
12.0	0.500000	0.451808	0.548192
16.0	0.500000	0.457190	0.542810
20.0	0.500000	0.461170	0.538830
24.0	0.500000	0.464146	0.535854
28.0	0.500000	0.466986	0.533014
32.0	0.500000	0.469171	0.530829
36.0	0.500000	0.470947	0.529053
40.0	0.500000	0.472436	0.527564
44.0	0.500000	0.473694	0.526306
48.0	0.500000	0.474792	0.525208
52.0	0.500000	0.475723	0.524277
56.0	0.500000	0.476572	0.523428
60.0	0.500000	0.477330	0.522670
	y_0	y_1	y_2
4.0	6.200000	6.774786	6.774786
8.0	7.689962	7.870938	7.870938
12.0	8.736837	8.882489	8.882489
16.0	9.598107	9.702190	9.702190
20.0	10.435579	10.531273	10.531273
24.0	11.071244	11.100038	11.100038
28.0	11.858994	11.956270	11.956270
32.0	12.458394	12.555744	12.555744
36.0	12.982830	13.079004	13.079004
40.0	13.427713	13.525402	13.525402
44.0	13.794744	13.894698	13.894698
48.0	14.080981	14.182261	14.182261
52.0	14.285610	14.392486	14.392486
56.0	14.413771	14.525860	14.525860
60.0	14.462750	14.560814	14.560814

switching is therefore needed for firms to undertake price competition.

According to our simulation, the frequency of switching can be sufficiently increased by simultaneous reduction in price. For example, Tables 1 and 2 show that α_1^* is increased quite moderately by a 10% reduction in p_1. Meanwhile, when both prices are reduced to 2.25 from 2.5, the effect on α_1^* is almost equivalent to a 35% increase in wealth, as is seen in Table 1 and Table 4. Simultaneous price reduction has a considerable magnitude of wealth effect in stimulating demand switching, hence increase in demand for both durable goods. Of course, these figures do not necessarily guarantee competitive price setting. There are many factors remaining to be considered. Nevertheless, at least it can be said that our specification is not incompatible with firms' price competition.

5. Discussion

In this paper we adopted the method of stochastic dynamic programming combined with expected utility theory. Such an infinite-horizon decision schemē has been criticized for its presumed too strong rationality and unrealistic computational capability.[5] We can, however, get qualitatively similar results without those presumptions. Our results remainvalid insofar as the consumer considers intertemporal allocation even in a very simple and primitive manner. We show an example in this section.

Suppose that the consumer determines her future consumption path in a manner proposed in Pemberton (1993). In other words, the consumer allocates her future expenditure equally into each future period subject to her budget constraint, which Pemberton calls sustainable consumption. The consumer need not worry about intricate details of intertemporal allocation among future periods. The consumer will be able to concentrate on the allocation between the current and the future instead, which is far easier and in some sense more realistic than to solve the complicated recursive functions of our model.

When the consumer has a unit of x_1, let y^* be the planned consumption of the perishable good in each future period without switching in the current period and y^{**} be that with switching. Further, let p^e be the expected expenditure on the durable goods in each future period. Then the present value of y^*+p^e or $y^{**}+p^e$ must be equal to w and discounted value of the income flow.

$$\frac{y^* + p^e}{r} = w + \frac{m}{r} - y_c \tag{38}$$

$$\frac{y^{**} + p^e}{r} = w + \frac{m}{r} - y_c - p_2 \tag{39}$$

In these equations y_c is the perishable good to be consumed in the current period. Generally p^e takes different values between the two cases, switched and unswitched. However, remember that the consumer does not adhere to precise computation in this example. The recurive function using this particular decision scheme can be written as follows.

$$V^*(w) = \max_{\alpha_1^*, y} H(\alpha_1^*)U(y_c) + \frac{\rho}{1-\rho}[\{1-F(\alpha_1^*)\}U(y^*) + F(\alpha_1^*)U(y^{**})] \quad (40)$$

Alhough this formulation is surely a kind of expected utility maximization, the basic ideas are completely different. The optimality conditions for y_c and α_1^* in this formulation are as follows.

$$H(\alpha_1^*)U'(y_c) = \frac{\rho r}{1-\rho}[\{1-F(\alpha_1^*)\}U'(y^*) + F(\alpha_1^*)U'(y^{**})] \quad (41)$$

$$\alpha_1^* U(y_c) + \frac{\rho}{1-\rho}U(y^*) = (1-\alpha_1^*)U(y_c) + \frac{\rho}{1-\rho}U(y^{**}) \quad (42)$$

There is only one difference from the original formulation, which is that $U(y^*)$ and $U(y^{**})$ have taken the places of $V_0(w_{+1})$ and $V_2(w_{+1}-p_2)$ respectively. It is obvious that $U(y^*)$ and $U(y^{**})$ are mathematically similar to those recursive functions, especially in concavity. The equations, (41) and (42), therefore satisfy all requirements to generate qualitatively similar results as derived in the third section.

Finally we have to discuss remaining problems. Our model clarifies inherently observable features of demand for durable goods generated by fluctuations in preference. Some important phenomena, however, remain unconsidered yet.

The durable goods considered in our model are supposed to be differentiated. Product differentiation is often accompanied by firm's nonprice strategies such as continuous model changes and advertisement. This fact means that firms have sufficient evidence that consumers' preferences are manipulatable or at least firms believe so. Consideration of firm's informational strategies is necessary to determine performance of markets for durable goods. Meanwhile, consumers' preferences interrelate to each other in various aspects, which correlates consumers' subjective probability distribution functions mutually. To aggregate individual demand functions explicitly it will be necessary to identify the correlation pattern as well as how $F(\alpha)$ diverges from the true distribution function. These remaining tasks are quite challenging while elaboration of the analytical framework to deal with quality of information is needed to tackle those problems.

Notes

1 The durable goods in our model are supposed to be accompanied by product differentiation and having no market for used goods, for example, clothing, furniture, tableware, electrical appliances and so on.

2 We admit that demand fluctuation, fashion or fads do not necessarily imply flexibility in preference. Those phenomena can happen even with fixed preference if consumers' expectations are affected by other persons' behavior. See, e.g., Bikchandani et al. (1993). Interdependent expectation, however, might be a kind of interdependent preference in broader sense. We also admit that demand switching does not always imply capricious behavior because technological restriction in consumption may force consumers to do so. Nevertheless, it is worthwhile to explore a model of flexible preference, for it has its own economic meanings.

3 As to an example of models of consumer's behavior applying the stochastic dynamic programming method, see Hey (1980) or Chapter 7 of Hey (1991).

4 With similar procedures one can examine effects of changes in r and ρ. It is obvious that r has similar effects to w. On the other hand, increase in ρ will reduce both of frequency of switching and the perishable good consumption in the current period. Meanwhile, ρ may be regarded as a flexible parameter that varies form period to period affected by other consumers' behavior. It would be also interesting to investigate interdependency among consumers' time preferences.

5 See, for instance, Hey (1983) and Pemberton (1993). Pemberton also proposes an alternative decision scheme. The example below is based on his intriguing idea.

References

Bikchandani, S., Hirshleifer, D., and Welch, I., (1993) A theory of fads, fashion, and cultural change as informational cascades, *Journal of Political Economy*, 100, 992-1026.

Hey, J.D., (1980) Optimal consumption under income uncertainty: an example and a conjecture, *Economics Letters*, 5, 129-133.

Hey, J.D., (1983) Whither uncertainty, *Economic Journal*, 93, 129-138.

Hey, J.D., (1991) *Experiments in Economics*, Blackwell, Oxford.

Leibenstein, H., (1950) Bandwagon, snob, and Veblen effect in the theory of consumers' demand, *Quarterly Journal of Economics*, 64, 183-207.

Pemberton, J., (1993) Attainable non-optimality or unattainable optimality: a new approach to stochastic life cycle problems, *Economic Journal*, 103, 1-20.

III. Recent Progess in Modeling Preferences and Risk Attitudes

DYNAMICALLY CONSISTENT PREFERENCES, QUADRATIC BELIEFS, AND CHOICE UNDER UNCERTAINTY

JÜRGEN EICHBERGER*
Department of Economics
Universität des Saarlandes
D-66041 Saarbrücken
Germany

SIMON GRANT*
Economics Programme (RSSS)
Australian National University
Canberra, ACT 6200
Australia

Abstract: In this paper we propose a new representation of preferences over uncertain acts than can accomodate both the Allais and Ellsberg paradox while retaining the form of representation under dynamically consistent updating. Moreover, this representation still allows one to distinguish risk preferences from beliefs.

1. Introduction

For almost fify years subjective expected utility theory has been the dominant paradigm for decision making under uncertainty, both from a normative and from a positive perspective. Though it became increasingly evident that actual decision makers quite commonly and consistently violate some of the tenets of the subjective expected utility hypothesis, it remains the most commonly used decision model in economics.

There are a number of reasons for the preference economists have for the expected utility model. First, the expected utility approach allows one to distinguish risk preferences from beliefs. One can separate the issue of how to deal with new information from the evaluation of outcomes. By updating the probability distribution that represents the beliefs of the decision maker in the usual Bayesian way, one obtains another expected utility functional with altered beliefs but unchanged risk preferences. Furthermore, the utility function over outcomes can be interpreted as characterizing a decision maker's

*We would like to thank Peter Wakker for helpful comments. Responsibility for any errors remains of course with us.

R. Nau et al. (eds.), Economic and Environmental Risk and Uncertainty, 195–205.

attitude towards risk. This utility function can be modified without affecting the beliefs incorporated in the probability distribution. Thus, one can analyze the effects of additional information and of changing risk preferences separately.

Secondly, expected utility theory has a strong normative appeal in the context of sequential decision making, if individuals can revise their actions after some part of uncertainty is resolved. A decision maker who satisfies the assumptions of the expected utility model will re-evaluate his choice independently of the unrealized consequences. This behaviour guarantees dynamic consistency of choice in the sense that decisions will be maintained conditionally on the new information.

Machina (1989) has pointed out that such a consistency requirement is not necessarily appropriate for preferences that do not conform to subjective expected utility hypothesis. Though in Machina's view dynamic consistency should be satisfied at any stage of a dynamic choice problem, as determined by the conditional preference relation, dynamic choice need not be independent of unrealized outcomes. However, studies of Sarin and Wakker (1992), Epstein and le Breton (1993) and Eichberger and Kelsey (1996) have shown that updating preferences in the dynamically consistent way, as suggested by Machina (1989) will often

- either imply that beliefs are represented by probability distributions
- or that the form of representation for the unconditional preferences will no longer hold for the updated preferences.

Sarin and Wakker (1993) dub the property that updated preference relations should also be represented by the family of functional that represents the unconditional preferences *sequential consistency.*

In this paper we characterize a family of preference relations over uncertain prospects that satisfy the following desiderata:

(i) preferences are dynamically consistent in the Machina sense and sequentially consistent;
(ii) beliefs can be separated from risk preferences;
(iii) preferences can accommodate Allais and Ellsberg type paradoxes.

The Allais and Ellsberg paradoxes are the most conspicuous violations of the subjective expected utility hypothesis. Finding a representation which can accommodate such consistently observed deviations from the expected utility model appears particularly desirable. For economic applications it is important that the representation allows us to separate risk preferences and beliefs and to remain within the same family of representation after updating in a dynamically consistent way.

The proposed representation builds on work by Chew, Epstein and Segal (1991) and Epstein and Segal (1992) who investigate quadratic utility functions under risk. Eichberger and Grant (1995) extend their approach to decision making under uncertainty by providing a set of axioms that imply the generalized representation. The quadratic functional obtained involves *two* probability measures defined on the set of events. Proper uncertainty aversion corresponds to these two probability measures differing. Lo (1995) has also observed that the quadratic expected utility model can be extended to choice under uncertainty and that it guarantees dynamically consistent choices without realizing its potential to separate risk preferences from beliefs.

In the next section, we introduce our model and quote the representation result obtained in Eichberger and Grant (1995). Section 3 shows how our model can handle the Allais and Ellsberg paradoxes and section 4 studies the portfolio choice problem as an economic application of our approach which promises some potential to resolve other inconsistencies between the subjective expected utility theory and observed behaviour. In particular, one can show that an uncertainty-averse decision maker may well choose a diversified portfolio, even if she is risk-neutral in regard to choice over lotteries.

2. The Model and General Results

Consider the Anscombe-Aumann setup for horse-race lotteries. $\mathcal{S} = \{s_1,...,s_S\}$ is a finite set of states and $\mathcal{E} = \{A,B,...\}$ denotes the set of events (the set of all subsets of $\mathcal{S}$). Let $\mathcal{X} = \{x,y,z,...\}$ be a finite set of (final) outcomes or consequences and $\Delta(\mathcal{X}) = \{P,Q,...\}$ the set of probability distributions *(lotteries)* on $\mathcal{X}$. The objects of choice are the mappings from $\mathcal{S}$ to $\Delta(\mathcal{X})$ *(horse-race lotteries)* in the set $\mathcal{H} = \{f,g,h,...\}$ which is ordered by a binary relation $\succsim$ representing the agent's preferences.

The set of horse-race lotteries $\mathcal{H}$ is a mixture space where, for all $\lambda \in [0,1]$ and all $f,g \in \mathcal{H}$, $\lambda f + (1-\lambda)g$ means the horse-race lottery that gives $\lambda f_s + (1-\lambda)g_s$ in state $s \in \mathcal{S}$. To simplify notation, we will write P for a horse-race lottery that gives the lottery prize P in each state and x for the degenerate lottery δ^x that yields the outcome x for certain. Furthermore, it will be convenient to describe by $f_E h$ the horse-race lottery which gives for each state s in event $E \in \mathcal{E}$ the lottery prize f_s and for states not in E the lottery prize h_s.

The following three axioms guarantee the existence of a non-zero function representing the horse-race lotteries.

Order: $\succsim$ is a complete pre-order.

Continuity:

For all $f,g,h \in \mathcal{H}$ such that $f \succ g \succ h$, the sets $\{\alpha \in [0,1] \| g \succ \alpha f + (1-\alpha)h\}$ and $\{\alpha \in [0,1] \| \alpha f + (1-\alpha)h \succ g\}$ are open.

Non-degeneracy: There exist acts $f,g \in \mathcal{H}$ such that $f \succ g$.

To get a further characterization, it is necessary to assume that preferences satisfy some form of monotonicity and independence. We shall require that preferences respect the partial ordering over lotteries of first order stochastic dominance, a weak form of monotonicity.

Ordinal State Independence:

For all outcomes $x,y \in \mathcal{X}$, all lotteries $P \in \Delta(\mathcal{X})$, all acts $h \in \mathcal{H}$, all $\lambda \in (0,1)$ and all non-null events $E \subseteq \mathcal{S}$,

$$x \succsim y \Leftrightarrow [\lambda \delta^x + (1-\lambda)P]_E h \succsim [\lambda \delta^y + (1-\lambda)P]_E h.$$

In expected utility theory beliefs over uncertain states are represented by probability distributions. Randomizing between two acts yields a new probability distribution over the same set of outcomes. Hence, one cannot have a decision-maker who is indifferent about two acts but strictly prefers a randomization (mixture) over these acts.

The Ellsberg paradox suggests that decision-makers dislike the uncertainty of a bet on states with unknown probabilities. If a decision-maker however perceives 'known' probabilities as more desirable than the 'unknown' probabilities over states, then she may well prefer an objective randomization over two indifferent acts, since it makes the expected outcomes in each state less extreme. Many notions of uncertainty aversion require therefore a preference for 'mixtures' of indifferent acts (e.g., Schmeidler 1989). We propose a particularly weak notion of uncertainty aversion called *Proper Uncertainty Aversion.* A preference for randomization is postulated only for conditional two-outcome lotteries.

***Proper Uncertainty Aversion*:**

For all two disjoint events $A, B \in \mathcal{E}$, and for all pairs of final outcomes x and y, with $x \succ y$, all $\lambda, p, q \in [0,1]$, with $p \succ q$ and all $h \in \mathcal{H}$

$$[\lambda P + (1-\lambda)Q]_A Q_B h \sim Q_A P_B h \text{ implies } \left[\left(\frac{\lambda}{1+\lambda}\right)P + \left(\frac{1}{1+\lambda}\right)Q\right]_{A\cup B} h \succsim Q_A P_B h$$

where $P = p\delta^x + (1-p)\,\delta^y$ and $Q = q\delta^x + (1-q)\delta^y$. Moreover, there exist two disjoint events A' and B' such that, if the corresponding first indifference holds, then the implication holds with strict preference.

If a decision maker's preferences satisfy proper uncertainty aversion, then the Ellsberg choices can be deduced. Notice that the existence of two ambiguous events is postulated so that the axiom is not vacuously satisfied.

The last axiom is a symmetry property which requires for any two indifferent acts that a preferred mixture has a symmetric mixture over which the decision maker is also indifferent. Hence, for any probability mixture which places more weight on f, there is another probability mixture which places more weight on g such that the individual is indifferent between the two randomizations. Applying this property to all pairs of indifferent horse-race lotteries provides us with the horse-race lottery analogue of Chew et al's (1991) *mixture symmetry axiom.*

Mixture Symmetry:

For every pair of acts $f,g \in \mathcal{H}$, $f \sim g$ implies: for all $\alpha \in (0,1/2)$ there exists $\beta \in (1/2,1)$ such that $\alpha f + (1-\alpha)g \sim \beta f + (1-\beta)g$.

One might reason further, that for each α the required β should be $(1-\alpha)$, that is „symmetric" mixtures of indifferent horse-race lotteries should themselves be indifferent. And indeed, as Chew et al (1991) have shown, given appropriate continuity and monotonicity assumptions for the preferences (Ordinal State Independence in our case) Mixture Symmetry is equivalent to the following axiom.

Strong Mixture Symmetry:
For every pair of acts $f,g \in \mathcal{H}$, and $\alpha \in [0,1]$,

$$f \sim g \text{ implies } \alpha f+(1-\alpha)g \sim (1-\alpha)f+\alpha g.$$

The following theorem follows as an almost immediate corollary of Chew et al's Theorem 5 (1991, p. 149).

Theorem 2.1: *Representation Theorem*

Let $\succsim$ satisfy Order, Continuity, Non-degeneracy, Ordinal State Independence and Proper Uncertainty Aversion. Then $\succsim$ satisfies Mixture Symmetry if and only if it can be represented numerically by a proper quadratic utility function V of the form

$$V(f) = \sum_{s\in\mathcal{S}} \sum_{t\in\mathcal{S}} \sum_{x\in\mathcal{X}} \sum_{y\in\mathcal{X}} \Psi(x,y,s,t)\, f_s(x) f_t(y). \tag{2.1}$$

Proof: Eichberger and Grant (1995). ■

The representation of preferences over acts (2.1) does not allow us to distinguish the decision maker's beliefs about states from the preferences over outcomes. Such a distinction is however desirable in many applications where one would like to fix risk preferences while analyzing the adjustment of beliefs to new information. Furthermore, it would be desirable to remain within the same class of representations after updating beliefs in a dynamically consistent fashion. Separability of risk preferences and beliefs together with the requirement that the representation be closed under dynamically consistent updating restricts the representation considerably.

Dynamic consistency requires that the representation in (2.1) be maintained after any event E that becomes known and for any act h previously chosen. The following lemma shows the restrictions on separability that follow from dynamically consistent updating.

Lemma 2.1: Suppose act h has been chosen and event E has been observed, then

$$V(f_E|E,h) = \sum_{s\in E} \sum_{t\in E} \sum_{x\in\mathcal{X}} \sum_{y\in\mathcal{X}} \hat{\Psi}(x,y,s,t|E,h) f_s(x) f_t(y).$$

is the *dynamically consistent* updated representation of $V(f)$ where

$$\hat{\Psi}(x,y,s,t|E,h) := \Psi(x,y,s,t) + \Psi_1(x,s|E,h) + \Psi_2(y,t|E,h) + \Psi_3(E,h)$$

with

$$\Psi_1(x,s|E,h) := [\sum_{t\notin E} \sum_{y\in\mathcal{X}} \Psi(x,y,s,t)\, h_t(y)] / |E|,$$

$$\Psi_2(y,t|E,h) := [\sum_{s\notin E} \sum_{x\in\mathcal{X}} \Psi(x,y,s,t)\, h_s(x)] / |E|, \quad \text{and}$$

$$\Psi_3(E,h) := [\sum_{s\notin E} \sum_{t\notin E} \sum_{x\in\mathcal{X}} \sum_{y\in\mathcal{X}} \Psi(x,y,s,t)\, h_s(x) h_t(y)] / |E|^2 .$$

Proof: Eichberger and Grant (1995, Lemma 3.1). ■

A dynamically consistent separation of risk preferences from beliefs must maintain the form of $\hat{\Psi}(x,y,s,t|E,h)$. In particular, the separation of Ψ must be such that beliefs and preferences over outcomes remain also separated in Ψ_1 and Ψ_2. As argued in Eichberger and Grant (1995),

$$\Psi(x,y,s,t) = \phi(x,y)\mu(s)\nu(t)) \tag{2.2}$$

is the only sensible separation of risk preferences from beliefs. It requires some straightforward but tedious calculations to show that dynamically consistent updates of this representation retain the form (2.2).

In (2.2), risk preferences are represented by the function ϕ which depends only on outcomes, and beliefs are represented by the functions μ and ν. One can normalize the belief functions μ and ν to satisfy

$$\sum_{s\in S} \mu(s) = \sum_{t\in S} \nu(t) = 1$$

so that one can interpret μ and ν as probability distributions. Uncertainty about the probability of a state is captured by combining these two probability distributions in a quadratic functional.

3. Ellsberg and Allais Effects

In this section we will show that the representation proposed in (2.1) and (2.2) is flexible enough to accommodate the Ellsberg and Allais paradox.

3.1 The Allais Paradox

The Allais paradox is one of the most famous inconsistencies between the expected utility hypothesis and actual choices of decision makers. In order to elicit the inconsistent behaviour, subjects are usually confronted with the following lotteries:

		Prizes		
		5	**1**	**0**
	L1	0	1	0
Lotteries	**L2**	0.1	0.89	0.01
	L3	0	0.11	0.89
	L4	0.1	0	0.9

When asked to choose between lotteries L1 and L2, most people preferred the certain prize of 1 to the gamble on 5 involved in lottery L2. These choices were particularly pronounced if prizes were very high, say 1 and 5 million dollars. The same subjects would however prefer L4 over L3 probably because, given that they had to take some risk, they preferred a 10 percent chance of winning 5 (million dollars) to the 11 percent chance of winning just 1. This contradicts the expected utility hypothesis since

$$V(\mathrm{L1}) := u(1) > 0.1 \cdot u(5) + 0.89 \cdot u(1) + 0.01 \cdot u(0) =: V(\mathrm{L2})$$

implies $\quad 0.11 \cdot u(1) > 0.1 \cdot u(5) + 0.01 \cdot u(0) \quad$ and

$$V(\mathrm{L4}) := 0.1 \cdot u(5) + 0.9 \cdot u(0) > 0.11 \cdot u(1) + 0.89 \cdot u(0) =: V(\mathrm{L3})$$

implies $\quad 0.1 \cdot u(5) + 0.01 \cdot u(0) > 0.11 \cdot u(1).$

The Allais paradox contradicts the fundamental property of expected utility theory that preferences are linear in probabilities. It is easy to see that the quadratic functional (2.2) can accommodate the observed behaviour. For example, let

$$\phi(x, y) = \frac{1}{2}[u(x)\mathrm{v}(y) + \mathrm{v}(x)u(y)]$$

with $u(5) = 100$, $u(1) = 99$, $u(0) = 10$ and $\mathrm{v}(5) = 100$, $\mathrm{v}(1) = 85$, $\mathrm{v}(0) = 0$. A straigthtforward calculation then yields (rounded to the next integer):

$$V(\mathrm{L1}) = 8415 > V(\mathrm{L2}) = 8411 \quad \text{and}$$

$$V(\mathrm{L3}) = 185 < V(\mathrm{L4}) = 190.$$

It is worth nothing that many other generalizations of the expected utility hypothesis, e.g., the Choquet integral of a capacity, maintain the linearity of preferences in beliefs and, therefore, cannot explain behaviour observed in the Allais paradox.

3.2 THE ELLSBERG PARADOX

Ellsberg (1961) challenged another tenet of expected utility theory, namely the view of Savage (1954) that individuals' beliefs about unknown states of nature can be adequately represented by additive probability distributions. To refute this hypothesis Ellsberg (1961) suggested the following test.

Consider an urn with 30 red balls and 60 balls that are either black or yellow. Four lotteries, given in the table below, were presented to a large number of people.

	Red	**Black**	**Yellow**
	30	**60**	
Lottery A	5	0	0
Lottery B	0	5	0
Lottery C	0	5	5
Lottery D	5	0	5

First decision makers had to choose between lottery A and lottery B and then between lottery C and D. Faced with the first choice, most people preferred lottery A over lottery B. When choosing between lottery C and D, however, lottery C was usually the preferred option. These decisions are blatantly inconsistent with Savage's *sure-thing principle*. Moreover, beliefs of decision makers with such preferences cannot be represented by an additive probability distribution:

- Since A is preferred to B, one is led to believe that the decision maker considers it less likely that a black ball is drawn than a red one, hence

$$\text{Prob}\,\{\text{black}\} < \text{Prob}\,\{\text{red}\} = 1/3.$$

- On the other hand, given that C is chosen rather than D, one concludes that the decision maker believes that red or yellow is more likely than black or yellow, i.e.,

$$\text{Prob}\,\{\text{red or yellow}\} < \text{Prob}\,\{\text{black or yellow}\}.$$

If probabilities are additive then

$$1/3 = \text{Prob}\,\{\text{red}\} < \text{Prob}\,\{\text{black}\}$$

follows. This contradicts the first assessment and therefore the existence of a subjective probability distribution that represents beliefs of such a decision maker. The Ellsberg paradox challenges not only Savage's theory but any theory of decision making under uncertainty based on *probabilistic sophistication* in the sense of Machina and Schmeidler (1992).

Once again, quadratic beliefs can accommodate the Ellsberg paradox. Assume, as before, that $\phi(x, y) = \frac{1}{2}[u(x)v(y) + v(x)u(y)]$ with u and v taking on the same values as they did in the previous subsection. Consider the following two probability distributions over states (R: „a red ball is drawn", B: „a black ball is drawn", Y: „a yellow ball is drawn"):

$$\mu(\text{R}) = 1/3, \quad \mu(\text{B}) = 1/3 - \varepsilon, \quad \mu(\text{Y}) = 1/3 + \varepsilon,$$
$$\nu(\text{R}) = 1/3, \quad \nu(\text{B}) = 1/3 + \varepsilon, \quad \nu(\text{Y}) = 1/3 - \varepsilon.$$

With these beliefs, the valuation of these lotteries becomes

$$V(\text{A}) = 4000/3 > V(\text{B}) = 4000/3 - 9000\cdot\varepsilon^2 \quad \text{and}$$
$$V(\text{C}) = 14000/3 > V(\text{D}) = 14000/3 - 9000\cdot\varepsilon^2.$$

This shows that the representation suggested in this paper can accommodate the Ellsberg paradox.

Note that this representation can deal with the Allais and the Ellsberg paradox while maintaining a distinction between risk preferences and beliefs[1]. The following section will study an economic example which shows that the quadratic representation suggested in this paper can capture stylized facts of economic behaviour that are difficult to model with expected utility theory.

---

[1] Sarin and Wakker (1992) note on page 1262 that their theory of Choquet expected utility can be extended to cover the case of rank-dependent expected utility on non-ambiguous events. It is well-known that rank-dependent expected utility can handle the Allais paradox.

4. An Economic Example: Portfolio Choice

This example will apply the quadratic belief approach to a portfolio choice problem. It can be shown that, in contrast to traditional expected utility theory, even a risk-neutral decision-maker may want to hold a mixed portfolio in order to diversify against uncertainty rather than risk. We choose risk-neutrality as the decision maker's risk attitude not because we believe that this is the most appropriate behavioural assumption, but to provide a simple illustration of the effect of uncertainty aversion on the standard portfolio choice problem.

Consider an investor who has to choose a portfolio (a,b) where

- a denotes the quantity of a risky asset which returns[2] $r(s)$ in state $s \in \mathcal{S}$ and
- b the quantity of a riskless asset returning $\bar{r}$ in each state.

The final wealth in state s of a portfolio (a,b) is

$$W(s|a,b) := r(s)\cdot a + \bar{r}\cdot b.$$

Using the same functional form as before to represent risk preferences

$$\phi(x,y) = \frac{1}{2}[u(x)v(y) + v(x)u(y)]$$

and given initial wealth W_O, the investor chooses the portfolio to maximize

$$V(a,b)$$

$$= \sum_{s\in\mathcal{S}} \sum_{t\in\mathcal{S}} \frac{1}{2}\cdot[u(W(s|a,b))\cdot v(W(t|a,b) + u(W(s|a,b))\cdot v(W(t|a,b)\cdot \mu(s)\cdot \nu(t)]$$

subject to the budget constraint $a+b \leq W_O$.

Assuming risk-neutrality, $u(x) = v(x) = x$, a straightforward calculation shows that

$$V(a,b) = [\sum_{s\in\mathcal{S}} r(s)\cdot \mu(s)\cdot a + \bar{r}\cdot b]\cdot[\sum_{t\in\mathcal{S}} r(t)\cdot \nu(t)\cdot a + \bar{r}\cdot b].$$

Notice that these preferences would coincide with the expected utility evaluation if $\mu(s) = \nu(s)$ would hold for all $s\in\mathcal{S}$.

Under the expected utility hypothesis a risk-neutral decision maker would evaluate a portfolio according to ist expected value given the unique probability distribution over states, $p(s)$, representing the decision maker's beliefs. Hence, traditional portfolio choice theory in the Savage tradition would predict that, for $\sum_{s\in\mathcal{S}} p(s)\cdot r(s) > \bar{r}$, the investor will sell the riskless asset in order to buy the risky asset and, for $\sum_{s\in\mathcal{S}} p(s)\cdot r(s) < \bar{r}$, she will sell the risky asset in order to buy more of the riskless asset. Without borrowing constraints unlimited arbitrage would make this problem indeterminate.

---

[2] These returns occur with certainty. The probability distribution that is associated with a state is concentrated on this outcome.

A risk-neutral investor with quadratic beliefs however will solve the following problem:

Choose a and b to maximize

$$[\sum_{s \in S} r(s) \cdot \mu(s) \cdot a + \bar{r} \cdot b] \cdot [\sum_{t \in S} r(t) \cdot \nu(t) \cdot a + \bar{r} \cdot b].$$

subject to $a + b \leq W_O$.

A straightforward calculation shows that

$$a^* = \frac{(\bar{r} - r_\mu) + (\bar{r} - r_\nu)}{(r_\mu - \bar{r}) \cdot (r_\nu - \bar{r})} \cdot \frac{\bar{r} \cdot W_O}{2}, \quad b^* = W_O - a^* \tag{4.1}$$

solves this problem, where $r_\mu := \sum_{s \in S} r(s) \cdot \mu(s)$ denotes the expected return of the risky asset evaluated with the probability distribution μ and $r_\nu := \sum_{t \in S} r(t) \cdot \nu(t)$ the expected return of the asset evaluated with the probability distribution ν.

The second order condition for a maximum is $2 \cdot (r_\mu - \bar{r}) \cdot (r_\nu - \bar{r}) < 0$. Thus, for the second-order condition to hold, we require either $r_\mu < \bar{r} < r_\nu$ or $r_\nu < \bar{r} < r_\mu$. That is, the uncertainty reflected in the two probability distributions μ and ν must yield expected returns on either side of the riskless return. If this condition on beliefs were not satisfied one asset would dominate the other even with uncertainty aversion. In particular, this condition will always fail to hold if there is no uncertainty aversion, that is, for $\mu = \nu$. In this case, $2 \cdot (r_\mu - \bar{r})^2 > 0$, if $r_\mu \neq \bar{r}$, and one asset dominates the other for a risk-neutral decision maker. In this case, the traditional analysis applies.

It is worth emphasizing that the investor's problem has a well-defined solution without any constraints on borrowing or lending. Moreover, from (4.1) we see that the investor sells the riskless asset in order to buy the risky asset, if and only if $\frac{r_\mu + r_\nu}{2} > \bar{r}$, that is, her „average“ belief about the expected return of the risky asset exceeds the riskless return.

Though the investor is risk-neutral in regard to objective risk, she is uncertainty-averse as is reflected by her two probability distributions μ and ν. Uncertainty aversion leads her to diversify even if she would never be willing to insure against objective risk if the price for insurance exceeds the expected loss.

In fact, it is easy to see that an uncertainty-averse but somewhat risk-loving decision maker with a quadratic representation would buy insurance against uncertain prospects and, at the same time, gamble in lotteries with objective risk.

5. Concluding Remarks

In this paper we proposed a new representation of preferences over uncertain acts that could accommodate the Allais and Ellsberg paradoxes while retaining the form of representation under dynamically consistent updating. Moreover, this representation still allows one to distinghish risk preferences from beliefs. This is particularly useful in economic applications where these notions have a well-established interpretation. Thus, one can model decision makers who are willing to gamble if they are faced with pure risk but who will be willing to buy insurance or to diversify portfolios if confronted by uncertainty about states. This provides one possible explanation for the old puzzle of people buying lottery tickets and at the same time insurance against accidents. It is worth noting that gambling, like playing roulette or buying a lottery ticket, often involves well-known objective probability distributions. The individual risk of an accident however usually cannot be associated with an unambiguous objective probability distribution.

References

Anscombe, F.J. and R.J. Aumann (1963), „A Definition of Subjective Probability", *Annals of Mathematical Statistics* 34, 149-205.

Chew, S.H., L.G. Epstein and U. Segal (1991), „Mixture Symmetry and Quadratic Utility", *Econometrica* 59, 139-164.

Eichberger, J. and S. Grant (1995), „Dynamically Consistent Preferences with Quadratic Beliefs", unbublished.

Eichberger, J. and D. Kelsey (1996), „Uncertainty Aversion and Dynamic Consistency", *International Economic Review*, forthcoming.

Ellsberg, D. (1961), „Risk, Ambiguity and the Savage Axioms", *Quarterly Journal of Economics* 75, 643-669.

Epstein, L.G. and M. le Breton (1993), „Dynamically Consistent Beliefs must be Bayesian", *Journal of Political Economy* 100, 691-712.

Epstein, L.G. and U. Segal (1992), „Quadratic Social Welfare Functions", *Journal of Political Economy* 100, 691-712.

Lo, IK.C. (1995), „Weighted and Quadratic Models of Choice under Uncertainty", unpublished, University of Toronto.

Machina, M. (1989), „Dynamic Consistency and Non-expected Utility Models of Choice under Uncertainty", *Journal of Economic Literature* 27, 1622-1668.

Machina, M.J. and D. Schmeidler (1992), „A More Robust Definition of Subjective Probability", *Econometrica* 60, 745-780.

Sarin, R.K. and P.P. Wakker (1994), „A General Result for Quantifying Beliefs", *Econometrica* 62, 683-686.

Sarin, R.K. and P.P. Wakker (1993), „Consistency in Dynamic Choice Situations with Nonexpected Utility Models", unpublished.

Sarin, R.K. and P.P. Wakker (1992), „An Axiomatization of Nonadditive Expected Utility", *Econometrica* 60, 1255-1272.

Savage, L. (1954), *The Foundations of Statistics*.John Wiley, New York.

Schmeidler, D. (1989), „Subjective Probability and Expected Utility without Additivity", *Econometrica* 57, 571-587.

INDUCED PREFERENCES AND DECISION-MAKING UNDER RISK AND UNCERTAINTY*

by David Kelsey, Department of Economics, The University of Birmingham

and Frank Milne, Department of Economics, Queens University, Canada

Abstract In this paper we suggest a new interpretation of non-additive probabilities. We study a decision-maker who follows the Savage axioms. We show the if (s)he is able to take unobservable actions which influence the probabilities of outcomes then it can appear to an outsider as if his/her subjective probabilities are non-additive. We make a related analysis of models with objective probabilities and show that the induced preferences can have the rank dependent expected utility form. Implications for multi-period decisions are explored.

1 INTRODUCTION

1.1 *Background*

The dominant theory of decision-making under uncertainty is subjective expected utility theory SEU (Savage 1954). One implication of this local risk-neutrality, (where the utility function is differentiable). Local risk-neutrality has some implausible implications, realist risk premia for small gambles imply excessively large risk premia for large gambles (Epstein 1992 p.11), individuals will never buy full insurance at an unfair premium (Segal and Spivak 1990) and investors will either buy or sell short *every* asset (Dow and Werlang 1992). There is also experimental evidence which fails to support SEU. The Ellsberg Paradox, (see Ellsberg 1961) contradicts implication (1).

Recent years have seen the development of some alternative decision theories, which do not imply local risk neutrality and are more compatible with the experimental evidence. Despite this, the theory which has the strongest normative justification is SEU. In the present paper we suggest a way to resolve this tension. We show that the decision-maker may at a fundamental level have SEU preferences. However to an outsider who is not able to observe all the relevant variables the induced preferences (Milne 1981) may appear to satisfy the axioms of one of the alternative theories. This theory of decision-making embodies the normative properties of SEU and the superior descriptive performance of the alternatives, at the same time.

Following the seminal work of Ellsberg (1961) there is a strong argument both on theoretical and empirical grounds that individuals often do not use conventional subjective probabilities, when making decisions under uncertainty. One of the more

R. Nau et al. (eds.), Economic and Environmental Risk and Uncertainty, 207–217.

prominent alternatives is that individuals have subjective probabilities which fail to be additive across disjoint events, Schmeidler (1989). The expected value of utility with respect to a non-additive probability distribution may be defined to be a Choquet integral, (Choquet 1953-4). The theory of maximising a Choquet integral of utility with respect to a non-additive probability will henceforth be referred to as *Choquet expected utility* (CEU). CEU is a generalisation of SEU.

CEU preferences may, under some assumptions, be given an alternative interpretation in terms of additive, but non-unique subjective probabilities, Gilboa and Schmeidler (1989), Kelsey (1994). In this case preferences may be represented as the minimum over a *set* of additive probabilities of the expected value of utility. This is known as *Maxmin Expected Utility theory (MMEU).* It embodies the intuition that when probabilities are not clear, individuals may simultaneously consider a number of probability distributions to be feasible. MMEU has the advantage that it can be expressed in terms of conventional expected values rather than the less familiar Choquet integral.

Dreze (1987) has modelled situations where an individual is able to influence the probabilities of the states by means of an (unmodelled) hidden action. He deduces that the individual has a convex set C of subjective probabilities. Any action is evaluated by the maximum value over the set C of expected utility and (s)he chooses the action which maximises the maximal value of expected utility. This can be motivated by assuming that the individual influences the probabilities to make them more favourable. Although the motivation is different, Dreze's axioms are similar to Gilboa and Schmeidler's (1989) axioms for (MMEU).

In this paper we explicitly model the hidden actions. We assume that the state space consists of both observable and unobservable states. We show that SEU preferences over the full state space will induce preferences of the CEU form over the observable states. Two hypothesis are used. First that the decision-maker is able to take an unobservable action, which makes the probability distribution more favourable. Second that nature may also take an unseen move which is unfavourable to the decision-maker. Of these, the first seems more plausible, however only the second is able to generate uncertainty-averse preferences. This is unfortunate since uncertainty-aversion is required to explain the Ellsberg paradox. Uncertainty-aversion appears to have more reasonable implications in some theoretical models, (see for instance Dow and Werlang 1992, Eichberger and Kelsey 1994). However as Kelsey and Quiggin (1992) note, preferences which are uncertainty loving (at least in some range) appear to be able to explain gambling behaviour better than SEU.

It could be that there are no actual moves by nature, but the perception of them causes individuals to act in an uncertainty-averse manner. It may be the case that the hidden actions are perceived to be available by the decision-maker but not by an outside observer. For instance, participants in a lottery may believe that choosing their "lucky number" will favourably influence the probability distribution.

1.2 *Updating Uncertainty Averse Preferences*

At present there is no generally accepted method of updating CEU preferences. This is a major barrier to applying CEU, since most economic problems, involve decisions taken at a number of points in time. The theory of induced preferences suggests a procedure for updating CEU preferences. It is generally agreed that SEU preferences should be updated by using Bayes' rule to revise the subjective probabilities. Hence when updating CEU preferences, which are induced by SEU preferences over some underlying space, it seems natural to take the Bayesian update of the underlying preferences. The new observable preferences are those induced by the updates of the underlying preferences. We show that under certain assumptions this procedure yields the same results as the commonly used Dempster-Shafer rule (Shafer 1976) for updating CEU preferences.

Organisation of Paper In the next section we introduce the CEU model. Section 3 shows how CEU preferences can arise as induced preferences when an individual is able to take a hidden action. In section 4 we study updating rules for induced preferences and show that they are dynamically consistent. Section 5 contains our conclusions.

2 DEFINITIONS

Notation 2.1 There is a finite set S of states of nature. A subset of S will be referred to as an *event*. The set of possible outcomes is denoted by X. An *action* is a function from S to X. The space of actions is denoted by $A(S)$. The decision-maker's preferences over $A(S)$ are denoted by $\succeq$.

Definition 2.1 A *non-additive probability (or capacity)* on S is a real-valued function ν on the subsets of S which satisfies the following properties

a. $A \subseteq B \Rightarrow \nu(A) \leq \nu(B)$ b. $\nu(\emptyset) = 0, \quad \nu(S) = 1.$

An expected value with respect to a capacity can be defined as a Choquet integral (Choquet 1953-4). We explain the Choquet integral below.

Notation 2.2 If $\phi: S \to \mathbb{R}$, let $\phi_{(i)}$ be the ith highest consequence of ϕ and $s_{(i)} = \phi^{-1}\phi_{(i)}$ be the state in which consequence $\phi_{(i)}$ occurs.

Definition 2.2 Choquet Integral If $\phi: S \to \mathbb{R}$, the Choquet integral of ϕ with respect to the capacity ν is defined by the following formula:

$$\int \phi(s) d\nu(s) = \phi_{(1)} \nu(s_{(1)}) + \sum_{i=2}^{n} \phi_{(i)} [\nu(s_{(1)} \ldots s_{(i)}) - \nu(s_{(1)} \ldots s_{(i-1)})].$$

Note that if ν is additive then this coincides with the usual definition of an expected value (for a discrete distribution).

Choquet Expected Utility (CEU) allows the decision-maker's subjective probability ν to be non-additive across disjoint events, ie. we do not require that $\nu(A \cup B) = \nu(A) + \nu(B)$, for $A,B \subseteq S$, $A \cup B = \emptyset$. CEU postulates that the decision maker has a utility function u and maximises the expected value of u with respect to the non-additive probability distribution ν, ie. $\int u(a(s)) d\nu(s)$.

3 INDUCED PREFERENCES

3.1 *The CEU Model*

In this section we model situations where the decision-maker can influence the probability of the states. Dreze (1987) has developed an axiomatic theory of decision-making under such circumstances. He shows that there exists a utility function $u: X \to \mathbb{R}$ and a convex set C of subjective probability distributions on S such that,

$$a \geq b \Leftrightarrow \max_{p \in C} E_p u(a) \geq \max_{p \in C} E_p u(b),$$

where $\mathbf{E}_p\, u(a)$ denotes the expected utility of action a with respect to probability distribution p. This is a special case of CEU.

Dreze assumes that the decision-maker is able to influence the probabilities by some process which is not modelled explicitly. We shall represent this by assuming that there are in addition unobservable states and actions. We show that if the individual has SEU preferences over pairs of observable and hidden actions, the induced preferences over the observable actions will be of the CEU form.

Let S_1 = first set of unobservable states,

S_2 = second set of unobservable states,

$\Sigma = S \cup S_1 \cup S_2$ = set of all states.

We model the decision process by assuming that, first the decision-maker chooses an observable action $a \in A(S)$. Let $r(a) \subset X$ denote the range of action $a \in A(S)$. Let $H_i(a)$ be the set of all functions from S_i to $r(a)$, for $i = 1,2$. (S)he is assumed to make a subsequent choice of a hidden action, $h_1 \in H_1(a)$. Nature then chooses a hidden action, $h_2 \in H_2(a)$. The restrictions on the choice of the hidden actions, ensure that their effect is to modify the probabilities of the consequences of the observable action. No new consequences are introduced by the hidden actions. In our opinion this represents the situation intended by Dreze. (A possible generalisation would be to restrict h_i to lie in a subset of H_i.) Clearly if new consequences could be introduced the presence of the hidden actions would eventually be revealed to the outside observer.

We shall assume that the decision maker has SEU preferences over the set of all triples $<a,h_1,h_2>$, $a \in A(S)$, $h_1 \in H_1$, $h_2 \in H_2$, with subjective probability π on Σ and utility function $u:X \to \mathbb{R}$. Such preferences could be given an axiomatic justification, (see, for instance Savage 1954).

Notation 3.1 If $a \in A(S)$ then the minimum of a, $m(a)$ (resp. the maximum of a, $M(a)$) will be defined by $m(a) = \min_{s \in S}\{u(a(s))\}$ (resp. $M(a) = \max_{s \in S}\{u(a(s))\}$).

It is clear that the decision-maker will choose his/her own hidden action to be a constant action which yields utility $M(a)$. Axiomatic models of decision-making when the payoff to the individual depends on a move by nature have been studied by Arrow and Hurwicz (1972), Barbera and Jackson (1988) and Barrett and Pattanaik (1995), among others. In these papers the decision-maker is assumed to have no beliefs over nature's move. Most of them conclude that the decision-maker must follow a "maximin-type" decision rule, which gives lexicographic priority to the worst outcome. We shall use these results to justify modelling a decision-maker who perceives nature's move to be the constant action which yields $m(a)$. This could also arise if the second hidden action were controlled by a "malevolent" individual who had interests diametrically opposed to those of the decision-maker. This would arise naturally in a two-person zero-sum game. These choices of hidden actions will yield induced preferences over observable actions which can be represented by the function,

$$V(a) = \sum_{s \in S} \pi_s u(a_s) + \delta_1 M(a) + \delta_2 m(a), \tag{3.1}$$

where $\delta_1 = \pi(S_1)$, $\delta_2 = \pi(S_2)$, π_s denotes the probability of state s, and a_s denotes the outcome which action a yields in state s. As the following proposition states the induced preferences over observable actions have the CEU form.

Proposition 3.1 Suppose that the underlying preferences are SEU with subjective probability π, then the induced preferences over $A(S)$ have the CEU form with the same utility function and capacity ν given by $\nu(A) = \pi(A) + \delta_1$, $(A \neq S)$, $\nu(S) = 1$.

Proof Let a be an action. Without loss of generality we may assume, $u(a(s_1)) > \ldots > u(a(s_n))$. If ν is defined as above:

$$E_\nu u(a) = (\pi_{s_1} + \delta_1)u(a_{s_1}) + \pi_{s_2}u(a_{s_2}) + \ldots + \pi_{s_{n-1}}u(a_{s_{n-1}}) + (\pi_{s_n} + \delta_2)u(a_{s_n})$$

$$= \sum_{s \in S} \pi_s u(a_s) + \delta_1 M(a) + \delta_2 m(a).$$

This is the same as equation 3.1, hence the Choquet integral does indeed represent the induced preferences.

■

3.2 *Rank Dependent Expected Utility*

So far we have interpreted π as a subjective probability on Σ. However our analysis would remain valid if π were an objective probability distribution. In this case, our arguments would show that an individual whose underlying preferences satisfy the von-Neumann Morgenstern axioms would appear to an outsider who could not see the hidden states, to be maximising the expected value of utility with respect to a distortion of the true probability distribution. Such preferences have been axiomatised by Quiggin (1982) and Yaari (1987) and are known as *rank dependent expected utility* (RDEU). We define RDEU preferences below.

Notation 3.2 If $a: S \to X$, let $a_{(i)}$ be the ith most preferred consequence of a and $p_{(i)}$ be the probability of consequence $a_{(i)}$.

RDEU implies that the decision-maker maximises,

$u(a_{(n)})g(p_{(n)}) + \sum_{i=1}^{n-1} u(a_{(i)})[g(p_{(i)}+...+p_{(n)}) - g(p_{(i+1)}+...+p_{(n)})]$, where g is a distortion of the probability distribution.

Suppose that the decision-maker maximises the expected value of the utility function u, with respect to the objective probability distribution π over the underlying state space Σ. To an outsider who can only observe the state space S, it will appear as if the probability is the distribution p on S, defined by $p_s = \pi_s/(\sum_{\sigma \in S} \pi_\sigma)$, where p_s is the probability of state s, as it appears to the outsider. The decision-maker will appear as if (s)he has RDEU preferences with distortion function given by $g(0) = 0$; $g(p) = \delta_2 + (1-\delta_1-\delta_2)p$, $0 < p < 1$; $g(1) = 1$. Apart from the discontinuities at 0 and 1, this function satisfies the properties postulated by Quiggin (1982), section 4. He argues that these preferences are able to explain both the Allais paradox and simultaneous gambling and insurance. Chew, Karni and Safra (1987) show that RDEU preferences will be risk-averse provided both u and g are concave. This implies that if u is concave the induced preferences will be risk-averse.

4 UPDATING

The CEU model has been successfully applied to a number of atemporal problems. (For a survey of applications see Kelsey and Quiggin 1992.) A major barrier to further applications is the lack of a generally accepted multi-period extension. For instance models of pricing derivative securities typically require trading in more than one time period. The theory of induced preferences suggests a solution to this problem. There is a strong case for using the Bayesian update of the underlying preferences, which will induce a set of updated preferences over the observable actions.

One difficulty with implementing this approach is that it is not clear how to apply Bayes' rule in this context. Suppose an event $E \subset S$, is observed. Then any subset $E' \subset \Sigma$ of the underlying state space is compatible with this observation provided, $E \subseteq E' \subseteq E \cup S_1 \cup S_2$. In other words, it is not clear after the observation of a subset E of the observable states, how many of the unobservable states will remain possible.

According to the most usual interpretation of CEU preferences, the non-observable states represent ambiguity in the decision-makers' beliefs. It would seem likely that as observations are made there would be a reduction in ambiguity. We shall model this in a simple way by assuming that after an observation fractions θ_1, θ_2 of S_1 and S_2 respectively, remain possible. (This is a special case of a proposal made by Gilboa and Schmeidler 1994). We formalise this in the following definition.

Definition 4.1 Suppose that the decision-maker has SEU preferences over actions on the underlying state space Σ with subjective probability π on Σ. Let θ denote the pair (θ_1, θ_2). The θ-Bayesian update π_E on Σ conditional on the observable event $E \sqsubset S$ is given by,

$$\pi_E(s) = \frac{\pi_s}{\sum_{t \in E} \pi_t + \theta_1\delta_1 + \theta_2\delta_2}, \quad \pi_E(S_1) = \frac{\theta_1\delta_1}{\sum_{t \in E} \pi_t + \theta_1\delta_1 + \theta_2\delta_2}$$

and $\pi_E(S_2) = \theta_2\delta_2/(\sum_{t \in E} \pi_t + \theta_1\delta_1 + \theta_2\delta_2)$.

Note that we do not require the same θ to apply each time preferences are updated.

Proposition 4.1 Under assumption 4.1 the induced preferences conditional on the observable event $E \subset S$ have the CEU form, with respect to the updated capacity ν_E defined by:

$$\nu_E(A) = \frac{\pi(A \cap E) + \theta_1\delta_1}{\pi(E) + \theta_1\delta_1 + \theta_2\delta_2}, \qquad A \cap E \neq E, \quad \nu_E(E) = 1.$$

Proof Note that the underlying preferences have the same general structure, before and after updating, hence it follows from Proposition 3.1 that the induced preferences have the CEU form. The form of the updated capacity ν_E follows from applying Proposition 3.1 to the updated probabilities π_E on the underlying state space.

■

Corollary 4.1 If $\theta_1 = 0$ and $\theta_2 = 1$, ν_E coincides with the Dempster-Shafer updating rule.

Proof In this case $\nu_E(A) = \dfrac{\pi(A \cap E)}{\pi(E) + \delta_1} = \dfrac{\nu((A \cap E) \cup \neg E) - \nu(\neg E)}{1 - \nu(\neg E)}$, which is the definition of the Dempster-Shafer rule.

Remark The assumption that $\theta_1 = 0$ and $\theta_2 = 1$, corresponds to a very pessimistic situation where the decision-maker believes that the only hidden action which has been taken is very unfavourable to him/her. The Dempster-Shafer rule, has been axiomatised by Gilboa and Schmeidler (1993), who also interpret it as a pessimistic updating rule.

5 CONCLUSION

5.1 *Related Literature*

There is some similarity between our results and those of Gilboa and Schmeidler (1994), however the motivation is different. They are presenting new representations of CEU preferences rather than finding functional forms for induced preferences. Like the present paper they show that CEU preferences are compatible with SEU preferences over an enlarged state space. They also consider updating and find sufficient conditions for the Dempster-Shafer rule. However they do not explicitly model the hidden actions.

5.2 *Extensions*

In Kelsey and Milne (1994) we argue that induced preferences will be dynamically consistent. In particular they will not be subject to Dutch books. Demonstrating dynamic consistency is important for the development of economic and financial applications of non-expected utility theories.

In this paper for the induced preferences to be uncertainty-averse we require that a hidden action be taken by a second individual or "nature". We assume that this second individuals is malevolent in the sense that (s)he acts to minimise the decision-maker's utility. This assumption of a malevolent individual is extreme. In Kelsey and Milne (1996) we show that it can be relaxed considerably. In that we paper we consider a model of incomplete information where the second individual has a number of possible types only one of which is malevolent. We show that the decision-maker will make uncertainty-averse choices *even if the probability of the malevolent type is arbitrarily small.*

ACKNOWLEDGEMENTS

Research supported by SSHRC (Canada) grant No. 410-92-0190 and the School of Social Science, The University of Birmingham. We would like to thank Richard Cornes, Jürgen Eichberger, Simon Grant, Stephen Morris, Clemmens Puppe, John Quiggin, and participants in the Royal Economic Society meetings Exeter 1994 and the FUR conference in Oslo 1994 for helpful comments.

REFERENCES

Arrow, K.J. and L. Hurwicz, (1972), in, "An Optimality Criterion for Decision-Making Under Ignorance." in C.F. Carter and J.L. Ford (eds.), *Uncertainty and Expectations in Economics*. Oxford, Blackwell.

Barbera, S. and M. Jackson, (1988), "Maximin, Leximin and the Protective Criterion." *Journal of Economic Theory,* vol. 46, pp. 34-44.

Barrett, C.R. and P.K. Pattanaik, (1995), "Decision Making Under Complete Uncertainty: The Ordinal Framework." in D. Dickinson and S. Sen (eds), *Uncertainty and Risk in Economic Life*, Edward Elgar, Aldershot, UK.

Chew, S.H., E. Karni and Z. Safra, (1987), "Risk Aversion and the Theory of Expected Utility with Rank Dependent Probabilities," *Journal of Economic Theory,* vol. 42, pp. 370-381.

Choquet, G., (1953-4), "Theory of Capacities", *Annales Institut Fourier*, vol. 5, pp. 131-295.

Dow, J. and S.R.C. Werlang, (1992), "Risk Aversion, Uncertainty Aversion and the Optimal Choice of Portfolio," *Econometrica,* vol. 60, pp. 197-204.

Dreze, J.H., (1987), "Decision Theory with Moral Hazard and State-Dependent Preferences." in J.H. Dreze, *Essays on Economic Decisions Under Uncertainty,* CUP, Cambridge.

Eichberger J. and D. Kelsey, (1994), "Non-Additive Beliefs and Game Theory," University of Birmingham discussion paper no. 94-01.

Eichberger J. and D. Kelsey, (1996), "Uncertainty Aversion and Dynamic Consistency." *International Economic Review*, forthcoming.

Ellsberg, D., (1961), "Risk, Ambiguity and the Savage Axioms," *Quarterly Journal of Economics,* vol. 75, pp. 643-669.

Epstein, L., (1992), "Behaviour Under Risk: Recent Developments in Theory and Applications." in J.J.Laffont (ed), *Advances in Economic Theory,* vol. II, CUP, Cambridge.

Gilboa, I. and D. Schmeidler, (1989), "Maxmin Expected Utility with a Non-Unique Prior," *Journal of Mathematical Economics,* vol. 18, pp. 141-153.

Gilboa, I. and D. Schmeidler, (1993), "Updating Ambiguous Beliefs," *Journal of Economic Theory,* vol. 59, pp. 33-49.

Gilboa, I. and D. Schmeidler, (1994), "Additive Representations of Non-additive Measures and the Choquet Integral," *Annals of Operations Research,* vol. 52, pp. 43-65.

Kelsey, D., (1994), "Maxmin Expected Utility and Weight of Evidence," *Oxford Economic Papers,* vol. 46, pp. 425-444.

Kelsey, D. and F. Milne, (1994), "Induced Preferences, Dynamic Consistency and Dutch Books," working paper, The University of Birmingham.

Kelsey, D. and F. Milne, (1996), "Induced Preferences, Non-Additive Probabilities and Multiple Priors," working paper, The University of Birmingham.

Kelsey, D. and J. Quiggin, (1992), "Theories of Choice Under Ignorance and Uncertainty," *Journal of Economic Surveys,* vol. 6, pp. 133-153.

Machina, M., (1989), "Dynamic Consistency and Non-Expected Utility Models of Choice Under Uncertainty", *Journal of Economic Literature*, vol. 27, pp. 1622-1668.

Milne, F., (1981), "Induced Preferences and the Theory of the Consumer," *Journal of Economic Theory,* vol. 24, pp. 205-213.

Quiggin, J., (1982), "A Theory of Anticipated Utility," *Journal of Economic Behaviour and Organisation,* vol. 3, pp. 323-43.

Savage, L.J., (1954), *Foundations of Statistics,* Wiley, New York.

Schmeidler, D., (1989), "Subjective Probability and Expected Utility without Additivity," *Econometrica,* vol. 57, pp. 571-587.

Segal, U. and A. Spivak, (1990), "First-order versus Second-Order Risk Aversion," *Journal of Economic Theory*, vol. 51, pp. 111-125.

Shafer, G. (1976), *A Mathematical Theory of Evidence*, Princeton University Press, New Jersey.

Yaari, M., (1987), "The Dual Theory of Choice Under Risk," *Econometrica,* vol. 55, pp. 95-115.

On Socks, Ties and Extended Outcomes

Harry M. Markowitz
Harry Markowitz Company
1010 Turquoise Street, Suite 245
San Diego, California 92109

As a result of conversations with Robin Pope during FUR VII, I now realize that the socks versus tie example of Markowitz 1959, Chapter 10, cannot be resolved in the way suggested there. I report below the additional problem, which I had not previously realized, and its resolution.

Chapter 10 describes examples which seem to be contradictions of the expected utility rule. It is argued there that they are not contradictions at all, but cases in which the set of outcomes has been misspecified. One example says that I would prefer a tie to socks for my birthday, but would not like to know which in advance. In other words, I would prefer a fifty-fifty chance of socks versus tie to knowing in advance which I will get.

This would seem to be a contradiction of the expected utility maxim since, with expected utility, if outcome A is preferred to outcome B, it is preferred to a fifty-fifty chance of A or B. The solution to this apparent contradiction is that "outcomes" here -- i.e., that which the decision maker seeks -- consists not only of socks or tie, but states of anticipation, surprise and the fun of the birthday ceremony.

To resolve the paradox it is sufficient to distinguish four outcomes depending on the decision maker's states in two time intervals:

$$0_1 = (KS, GS)$$
$$0_2 = (KT, GT)$$
$$0_3 = (FF, GS)$$
$$0_4 = (FF, GT)$$

R. Nau et al. (eds.), Economic and Environmental Risk and Uncertainty, 219–226.

where:	is short for:
KS	"know socks"
GS	"get socks"
KT	"know tie"
GT	"get tie"
FF	have a fifty-fifty chance of socks or tie.

GS represents both the receipt of socks and their subsequent use. GT, KS, KT and FF similarly may represent a sequence of states during one of the two time intervals distinguished here. The various states during the time interval need not be separately denoted for this analysis.

Since there are four outcomes -- 0_1, 0_2, 0_3, 0_4 -- the Von Neumann and Morgenstern (VN-M) axioms[1] assume, among other things, that the decision maker has a preference ordering among probability distributions of these outcomes: (p_1, p_2, p_3, p_4) where p_i is the probability of the i-th outcome. The VN-M axioms imply that there exist numbers u_1, u_2, u_3, u_4 such that $P = (p_1, p_2, p_3, p_4)$ is preferred to $Q = (q_1, q_2, q_3, q_4)$ if and only if

$$\Sigma u_i p_i > \Sigma u_i q. \tag{1}$$

With outcomes defined as above, the decision maker's preferences do not contradict expected utility. They only require that

$$1/2 u_3 + 1/2 u_4 > u_2 > u_1 \tag{2}$$

It is usually assumed that the decision maker can rank any two probability distributions P and Q. There is a problem with this not recognized in Markowitz (1959). Consider, for example, the probability distribution (0, 0, .3, .7). This is a thirty percent chance that the probability is fifty-fifty, and socks result; and a seventy percent chance that the probability is fifty-fifty, and a tie results. Ranking this probability distribution makes no more sense than ranking a fifty-fifty chance that two equals one.

It follows that the only probability distributions which can be contemplated, much less ranked, are those of the form $(p_1, p_2, .5(1-p_1-p_2), .5(1-p_1-p_2))$[2].

The probability distributions P which are of this form constitute a convex set S. One can postulate the axioms of expected utility for probability distributions P, Q, R... from the set S. In this case there exist u_1, u_2, u_3, u_4 which order P, Q, R... according to preferences, but these u_i are not unique. As usual, the origin and scale of utility is arbitrary; e.g. since 0_2 is preferred to 0_1, we may arbitrarily assign $u_1 = 0$, $u_2 = 1$. Usually, once a zero and unit are thus decided, all other u_i are uniquely determined. In the present case, if (u_1, u_2, u_3, u_4) describe preferences then so do $u_1, u_2, u_3 + c, u_4 - c$ for any real number c; since

$$u_1p_1 + u_2p_2 + .5(1\text{-}p_1\text{-}p_2)u_3 + .5(1\text{-}p_1\text{-}p_2)u_4$$
$$= u_1p_1 + u_2p_2 + .5(1\text{-}p_1\text{-}p_2)(u_3\text{+}c) + .5(1\text{-}p_1\text{-}p_2)(u_4\text{-}c) \qquad (3)$$

One way to resolve the ambiguity in choice of c is to assume that the utility of the game is the sum of the contribution from the having of extra socks, having a new tie, and playing the birthday game. Thus,

$$1/2U(FF,GS) + 1/2U(FF,GT) = U(G) + 1/2U(KS,GS) + 1/2U(KT,GT) \qquad (4a)$$

$$U(FF,GS) = U(G) + U(KS,GS) \qquad (4b)$$

$$U(FF,GT) + U(G) + U(KT,GT) \qquad (4c)$$

Here we write U(KS,GS) for u_1, etc.; and U(G) is the utility of playing the birthday game. For example, if we let $u_1 = 0$, $u_2 = 1$ by convention, and if $1/2u_3 + 1/2u_4 = 2$ -- because the decision maker is indifferent between 0_2 with certainty (i.e., (0,1,0,0)) and a 50-50 chance of $0_1 = (1,0,0,0)$ versus (0,0,1/2,1/2) -- then

$$2 = U(G) + 1/2; \text{ i.e., } U(G) = 1.5 \qquad (5a)$$

$$u_3 = 1.5 + 0 = 1.5 \qquad (5b)$$

$$u_4 = 1.5 + 1 = 2.5 \qquad (5c)$$

Therefore $1/2u_3 + 1/2u_4 = 2.0$ as required.

Extensions and Reflections

1. As John Pratt explained during the Roundtable, the above remarks readily generalize to two or more chance situations. These chance situations may be (a) alternates, or (b) offered at different times. An example of "(a) alternates" would be either
   (i) I play the slot machines, or
   (ii) I play roulette.

To simplify the example, I don't have time for both. An example of "(b) chance situations offered at different times" would be one in which one strategy I could follow is to
   (1) play roulette on Saturday (and follow a particular betting rule), then
   (2) go to the races on Sunday (and follow a particular betting rule there).

An outcome for the latter situation would, as before, be a sequence of states during different time intervals. For example, one outcome might be:

> (spend Saturday playing roulette winning/losing x dollars,
> spend Sunday at the tracks winning/losing y dollars,
> therefore wake up Monday with z dollars in pocket.)

Such (a) alternate gambles and/or (b) successive gambles can be nicely represented by the tree diagrams which John Pratt presented at the Roundtable. For ease of presentation, John assumed that one sure outcome had higher utility and one had lower utility than all gambles. This assumption is not necessary, as illustrated in the example at the end of the previous section.

2. The above argument suggests that outcomes can always be elaborated to fit gambling preferences. Robin Pope is not convinced. I don't want to argue that matter right here. Rather, I want to report a private discussion (on the bus from hotel to conference) between Mark Machina and me.

Mark argued that even assuming that outcomes can always be elaborated to represent gambling (i.e., risk situation) preferences, it may not always be most convenient to do so. The analysis may be "neater" in a space of outcomes where expected utility does not apply than in a more complex one where it does. I agreed with Mark, and still do.[3]

3. Let's return to an example of a single chance situation, like "socks versus tie". I would prefer to see the home team win than lose; but I would prefer not to know in advance, before I go out to the ballpark. A game in which the home team was so strong and the visitors so weak that the former was sure to win, would be little fun to watch.

The resolution of this "paradox" is as in the "socks and tie" example. As in that example, we can solve for the utility of the game itself U(G), which here I write as $U(G_{.5})$ for a game with a .5 chance. But it is not necessary that we will get the same U(G) from formula (4) from a 90-10 game as we do from a 50-50 game (using the appropriate probabilities in (4)). Perhaps a 90-10 game is duller than a 50-50 game. Thus $U(G_p)$ which solves

$$pU(G_p,W) + (1\text{-}p)U(G_p,L) = U(G_p) + pU(W) + (1\text{-}p)U(L)$$

may be a function of p (where U(W) is utility of a win, $U(G_p,W)$ is the utility of watching a game with probability p of winning and seeing the home team win, etc.).

Various views may be illustrated by this example:

(a) $U(G_p,W)$ and $U(G_p,L)$ are, I believe, what R. Pope refers to as probability dependent utilities.

(b) One may amplify the definition of outcome so that EU still holds.

Then outcomes include, for example:

$$(G_{.3}, W)$$

$$(G_{.3}, L)$$

$$(G_{.4}, W)$$

etc.

If, for simplicity, we only allow a finite number of values of p, then there are a finite number of outcomes which can be labeled $0_1, \ldots, 0_n$ and analyzed as above. As before, only a convex subset of probability distributions $(p_1, \ldots, p_n)$ are thinkable. The situation can be illustrated by J. Pratt's trees.

If we allow any p in the interval $0 \leq p \leq 1.0$ then we have to use a different notation, sprinkle the word "measurable" here and there like holy water, but nothing essential changes.

(c) Finally, perhaps there exists a simpler space than the set of all

$$\left(G_p, \begin{Bmatrix} W \\ L \end{Bmatrix} \right)$$

such that preferences can be easily described but are not EU in the simpler space.

Appendix
Axioms for Multiperiod Analysis

Markowitz (1959), Chapter 11 on "Utility Analysis Over Time" specifies that: "An *outcome*, in the present discussion, is a time pattern of consumption $(C_1, C_2, \ldots, C_T)$. If the C_t represent amounts of money devoted to consumption, an outcome is a 'history' of consumption expenditures. If the C_t are vectors of goods and services enjoyed, then consumption is a time series of such enjoyments". (The inclusion of a bequest value at the end of the outcome vector makes no essential change to the analysis.) Time series of enjoyments postulated as outcomes in the body of the present paper are special cases of Markowitz (1959) Chapter 11 outcomes.

For convenience, Chapter 11 assumes that there are only a finite number of possible outcomes thus defined; e.g., a finite number of possible consumption numbers or vectors C_t as of any time t and a finite (T) number of periods, hence a finite number of outcomes, i.e., time series

$$O_j = (C_{1j}, C_{2j}, \ldots, C_{tj}) \quad j = 1, \ldots, n. \tag{6}$$

In other words "we can therefore label possible outcomes with numbers 1, 2, . . ., j, . . ., n." Then a probability distribution of possible outcomes -- i.e. a probability distribution over all possible histories of consumption -- is a vector

$$(p_1, \ldots, p_n)$$

(where, of course, $p_j \geq 0$ and $\Sigma p_j = 1$).

The remainder of Chapter 11 notes that the formal axioms which were applied in Chapter 10 to probability distributions of outcomes at one point in time could also be asserted for probability distributions of Chapter 11 types of outcomes; considers "whether formal relationships expressed in the axioms appear as plausible when applied to probability distributions of outcomes spread over time as they do when applied to outcomes resulting from single period choice situations"; and points out that (obviously) if preferences among the probability vectors $P = (p_1, \ldots, p_n)$ in Chapter 11 satisfy the same axioms as the probability distributions $P = (p_1, \ldots, p_n)$ in Chapter 10, then they too can be ordered by attaching a number u_i to outcome

$$O_i = (C_{1i}, \ldots, C_{Ti})$$

and choosing P to $Q = (q_1, \ldots, q_n)$

if and only if

$$\Sigma p_i u_i > \Sigma q_i u_i.$$

Chapter 11 uses Bellman's dynamic programming arguments to reduce the many-period choice of strategy to a sequence of single period expected utility maximizations. Chapter 11 then discusses other matters related to the u_i and dynamic expected utility analysis; e.g. it points out that u_i is <u>not necessarily</u> of the form $u_i = \Sigma U_t(C_{ti})$, etc.

It was my memory, or assumption, as of August 1, 1994, that the treatment of an "outcome" as a time series as in (6) was standard by 1959. I looked in Von Neumann and Morgenstern (1967[4]) so that I could cite them as considering their axioms applicable to the four outcomes in the socks versus tie example. However, in their discussion of expected utility, Von Neumann and Morgenstern state in Section 3.3.3, that "it would be an unnecessary complication, as far as our present objectives are concerned, to get entangled with the problems of the preferences between events in different periods in the future. It seems, however, that such difficulties can be obviated by locating all 'events' in which we are interested at one and the same, standardized moment, preferably in the immediate future". Thus, in their discussion of expected utility, Von Neumann and Morgenstern deal only with the single period case. (Most of their book is concerned with games which may have many moves, but have a single payoff to each player at the end of the game.)

I am somewhat at a loss as to who to cite on this matter. The examples by Bellman that I looked at treat the dynamics of games, like those of VN-M, with a single scalar payoff. Perhaps the Savage (1954) discussion of Small Worlds (Section 5.5) is the proper reference. Markowitz (1959) Chapter 11 may be considered a spelling out of the finite case of the matter as Savage discusses it there.

References

Bellman, R. E. (1957) *Dynamic Programming*, Princeton University Press.

Markowitz, H. M. (1959) *Portfolio Selection: Efficient Diversification of Investments*, Wiley, Yale University Press, 1970, Basil Blackwell, 1991.

Savage, L. J. (1954) *The Foundations of Statistics*, Wiley, 2nd edition, Dover, 1972.

Von Neumann, J. and Morgenstern, O. (1944) *Theory of Games and Economic Behavior*, Princeton University Press, 3rd edition, Wiley, 1967.

Endnotes

[1]At least the version of the VN-M axioms used in Chapter 11 of Markowitz (1959) allows "outcomes" to be sequences and assumes that the decision maker orders probability distributions of outcomes thus defined. See the Appendix on "Axioms for Multiperiod Analysis".

[2] This assumes that the decision maker can choose randomly between 0_1, 0_2 and a 50-50 chance of 0_3, 0_4, with negligible emotional effect of this randomization. For example, perhaps this emotionally neutral randomization consists of typing values for p_1 and p_2

into a computer, pressing enter, and the random choice appears immediately on the screen.

[3]On rereading Markowitz (1959) I find that my view then (p. 225) was that "the expected utility maxim can be extended to include such considerations; but a large number of such 'extensions' transform the maxim from a convenient rule to a useless formality."

[4]The preface to the second edition dated September 1946, says that "we have added an Appendix containing an axiomatic derivation of numerical utility. This subject was discussed in considerable detail [in the first edition], but in the main qualitatively, in Section 3."

HIERARCHIES AND THE SELF-CONTROL OF RISK ATTITUDE

LUC WATHIEU[1]
The Hong Kong University of Science and Technology
Clear Water Bay, Kowloon, Hong Kong

Abstract. In hierarchies, top members are not continuously informed about the intermediary outcomes of their policies. This feature of organizational decision making is often regarded as a drawback caused by communication costs. On the contrary, we argue that such information decentralization creates a valuable interface between a decision maker and his environment. We assume a reference-dependent utility function à la Kahneman and Tversky. Personal contact with an intermediary outcome provokes a shift of reference point, and thereby a change of attitude towards the forthcoming alternative risks. Information decentralization stabilizes the reference point. We characterize classes of dynamic choice situations where this effect should prove desirable. Then, we show that hierarchical systems can help decision makers overcome their personal conservative tendencies.

1. Introduction

Chandler (1962) implied that, by decentralizing an organization, one can modify the psychological attitude of the involved decision makers:

"The basic reason for its [the M-form of organization] success was simply that it clearly *removed the executives* responsible for the destiny of the entire enterprise from the more routine operational activities, and so gave them the time, information and even *psychological commitment for long-term planning and appraisal.*"[2]

The nature and implications of the relationship between decentralization and "psychological commitment" has received no attention from the decision making literature. However, these issues are crucial to understand both decision making in organizations and the organization of decision making. This chapter makes a formal investigation of the link between decentralization and a decision maker's attitude in a dynamic choice context under risk.

---

[1] Thanks are due to Bernard Sinclair-Desgagné, Spyros Makridakis, Anil Gaba, Robin Hogarth, Robert Winkler, Saugata Banerjee, Itzhak Gilboa, Jerry Ross, and seminar audiences at INSEAD, HKUST, FUR VII and EIASM. Financial support from ICM, the Wei Lun Fund and INSEAD is gratefully acknowledged.

[2] p. 309, italic added.

R. Nau et al. (eds.), Economic and Environmental Risk and Uncertainty, 227–243.

1.1. BASIC INTUITION

It is a well-documented fact that decision makers use reference points when they evaluate the possible outcomes of risky monetary alternatives. A reference point is a benchmark that determines whether an outcome is perceived as a loss or as a gain. Evaluation of losses and gains is typically asymmetric, since loss aversion is a frequent and dominant feature of human preferences (see, e.g., Tversky and Kahneman, 1991). The most common reference point, according to the literature, is accumulated wealth.

In a dynamic context, as wealth evolves, the reference point shifts and the utility associated with each wealth position changes. For instance, a position that initially corresponds to a gain can be evaluated as a loss after an increase in wealth has been perceived. The present chapter examines the problem of dynamic choice under risk for an expected utility maximizer whose reference point is subject to such volatility.

Suppose that you lose $1000 on one of your personal investments. Loss aversion predicts that you experience a severe dissatisfaction. Shortly after, this investment yields a gain of $1000. Now, when you consider your experienced utility over the period, you may find that you would have preferred to hear nothing about the intermediary outcome and be directly exposed to a cumulated payoff of $0. When such is the case, it means that you did not experience the $1000 gain as a compensation for the earlier loss. This phenomenon corresponds to a shift of reference point. As this example illustrates, an individual with a volatile reference point should develop, ex ante, a preference relation over the frequency of feedback. Indeed, when such frequency is under control, it is possible to stabilize the reference point and the corresponding utility function for a while. Thus a meta-decision can be made, that determines future preferences among alternative risks.

Feedback, however, is not just stimulus for (dis)satisfaction and reference point shifts. It often conveys useful information about future alternatives. How, then, could one be able to process any such information while controlling exposure to intermediary outcomes? *The necessary device is hierarchical.* In hierarchies, decision makers choose policies, order their implementation and organize a calendar of appraisals. Subordinates are then in charge of processing information, executing the policy and reporting in due time. Because they allow temporary *information decentralization,*[3] hierarchies permit to differentiate between processed information and perceived information. Perceived information anticipatively motivates policy making at the top, whereas processed information permits contingent implementation at the lower level.

Thus our theory is that, by decentralizing information about the intermediary outcomes of his plans, a decision maker can (and often should) consistently achieve self-control of his risk attitude, while remaining capable of contingent planning. This chapter is devoted to an analysis of the circumstances under which information decentralization and the implied stabilization of the reference point are desirable, in

[3] We borrow this term from Marschak and Radner (1972). When we write "(de)centralization", we always refer to "information (de)centralization".

terms of the decision maker's expected utility. We will also examine some properties of the resulting choice behavior, and show that decentralization is a remedy against individual conservatism in a number of situations.[4]

1.2. LINK WITH ORGANIZATION THEORY

That top members of hierarchies are not kept continuously informed about the intermediary outcomes of their policies is often regarded as an unavoidable drawback by organization theorists. For instance, Marschak and Radner (1972) relate the need for decentralization to communication costs. Here, in contrast, we argue that information decentralization can create a valuable interface between decision makers and their environment, which enables them to consistently adopt a desired risk attitude.

In that sense, we share the viewpoint originally expressed by Barnard (1968), according to which the "essence of formal organization" is to achieve "deliberate adoption of means and ends" that are out of reach for isolated individuals. However, we do not deal with the limitations of human capabilities (as Radner, 1992, or Geanakoplos and Milgrom, 1991, for instance). Rather, we focus on the peculiarities of individual risk attitudes that are motivated by what Tversky and Kahneman (1991) call "the psychophysics of hedonic experience."

If reference point shifts are considered as "biases", what we claim here is that hierarchies are debiasing systems (to be added to the list of Arkes, 1991), that decision makers should find attractive under some circumstances. Postrel and Rumelt (1992) have a similar philosophy, when they suggest that hierarchies offer a remedy to individual impulsiveness by imposing routines and habits on members.

Our emphasis on the transformation of risk attitude in hierarchies could be related to the observation by McCrimmon and Wehrung (1986) that decision makers are less risk averse in business context than in personal context.

1.3. LINK WITH DYNAMIC CHOICE THEORY

Our work can also be seen as an essay in dynamic choice under changing tastes. This field of research, that was initiated by Strotz (1956), primarily referred to addictive consumption goods such as tobacco, alcohol or food. Only recently, with a debate on the dynamic implications of non-expected utility theories, have we seen the appearance of a more general concern about time-inconsistent preferences (see, e.g., Machina, 1989). Strotz and his followers pointed the structure of time preference (i.e., non-exponential discounting) as the key factor to explain the changing taste phenomenon. The possible role of reference point shifts was only introduced lately, by Loewenstein (1988).

How to rationally decide under changing preferences is still largely considered as a puzzle (McClennen, 1990). Information decentralization is here offered as a new

---

[4] On conservatism in individual decision making, see Samuelson and Zeckhauser (1988).

solution for non-myopic decision making under changing utility of wealth, when change is caused by reference point shifts. Interestingly, this solution goes beyond the two traditional solutions of precommitment and sophisticated choice.

Decentralization resembles precommitment (see, e.g., Schelling, 1978, and Elster, 1979), because it extends the scope of the initial viewpoint (and of the resulting preferences) over the future. But it does not imply any kind of regret in the future, nor the canceling of any option. Precommitment alters future alternatives, whereas decentralization stabilizes preferences. In a way, we follow Stigler and Becker (1977), who argue that changing preferences always correspond to a stationary pattern, when the determinants of change are understood. Our work identifies one such determinant (reference point shifts caused by outcome perception) and makes it a controllable variable.

We actually use a "sophisticated" approach as a starting point, i.e., we regard future choices as exogenous constraints for current preferences. But we also assume that the decision maker has initially access to an information system that can determine his future risk attitude. Thus, we end up with a decision procedure where initial choices constrain future preferences. This contrasts with the classical sophisticated solution, which recommends resignation. In that sense, information decentralization in a context of reference point shifts provides an operational device for implementing McClennen's (1990) recent suggestion of "resolute choice" as a solution for planning under changing tastes ("resolute choice" occurs when the future self decides to implement a plan that was elaborated according to initial preferences).

An important aspect of our theory is the distinction between two effects of feedback information: it provokes reference point shifts, and it dictates the implementation of the optimal contingent plan. Information decentralization in a hierarchical context dissociates the two effects (which would otherwise be simultaneous, for an individual decision maker). This observation allows us to give an original insight about preferences over the timing of uncertainty resolution (in comparison with, e.g., Kreps and Porteus, 1978, or Bell, 1985). Indeed, in a hierarchical context, an individual at the top may prefer to delay personal contact with information, while always ordering its immediate processing by a subordinate.

1.4. CLOSELY RELATED WORK

Two papers inspired by behavioral decision theory have some similarities with the present essay. Thaler (1985) examines whether a given sequence (x, y) of outcomes gives greater satisfaction when "integrated" (i.e., coded as a sum $x + y$) than when "segregated" (i.e., coded as a sequence of distinct stimuli). In our terms, "integration" corresponds to information decentralization, while "segregation" results from permanently tracking the outcomes. Our approach is different because we analyze the ex ante decision to decentralize information, in a context of uncertainty and learning. Also, we study how a decision to decentralize impacts the chosen policies, an aspect that Thaler ignored.

Kahneman and Lovallo (1993) also discuss loss aversion in an organizational context. But they do not view organizations as systems where preferences are engineered (through controlled perception) so as to achieve some desirable decision behavior. They rather conclude that "the forces that produce risk aversion in experimental studies of individual choice may even be stronger in the managerial context." However, after one of their numerical examples, they mention that "reducing the frequency of evaluations can mitigate the inhibiting effects of loss aversion on risk taking" (p. 23), which gives a hint for the present chapter.

After having set up the model in the next section, we give a characterization of dynamic choice situations where decentralization of information is optimal (assuming it is freely available). Then, in section 4, we perform the reverse analysis to find situations where centralization (i.e., the decision maker permanently keeps track of the outcomes) should be preferred. In section 5, we show that information decentralization constitutes a device to overcome conservatism, in two important classes of problems. Section 6 comments on a side effect of information decentralization: it creates a risk of escalation. Section 7 concludes.

2. Model

2.1. DYNAMIC CHOICE SITUATION

We consider the problem of a decision maker facing a two-period choice situation. There is a finite set X of possible monetary outcomes. A representative outcome $x \in X$ is denoted by x_1 when it results from a first period action and by x_2 when it results from a second period action. The set of initially available actions is denoted by A_1, with representative element a_1. Each action a_1 is a probability measure defined on a Boolean algebra $\mathbb{B}$ of subsets of X. Both initial action and the obtained first period outcome may determine the performance of period 2 actions (e.g., in case of learning). The choice set in the second period is therefore denoted by $A_2(a_1, x_1)$, with representative element $a_2(.|a_1, x_1)$, a probability measure defined on $\mathbb{B}$.

2.2. UTILITY FUNCTION

Outcomes are evaluated through an increasing utility function that has the basic features of Kahneman and Tversky's (1979) value function. We assume that accumulated wealth serves as a reference point (as initially suggested by Markowitz, 1952), and the received outcomes are direct stimuli for (dis)satisfaction. We moreover assume loss aversion and a weak form of diminishing sensitivity. These assumptions correspond to realistic features of human "psychophysics of hedonic experience". Empirical support under the von Neumann-Morgenstern framework can be found in Fishburn and Kochenberger (1979), for instance.

Formally, we consider a utility function $v{:}\bigcup(X,-X,X\oplus X,\{0\})\to I\!R$ that is characterized as follows:[5]

$$v(0)=0, \tag{1}$$

$$v(x)<v(y) \quad for\ x<y, \tag{2}$$

$$v(x+y)-v(x)\le v(y) \quad for\ x,y\ge 0 \text{ (subadditivity for gains)}, \tag{3}$$

$$v(x+y)-v(x)\ge v(y) \quad for\ x,y\le 0 \text{ (superadditivity for losses)}, \tag{4}$$

$$v(x)<-v(-x) \quad \text{(loss aversion)}. \tag{5}$$

In order to focus on the role of loss aversion, we will sometimes work with a special form of v that we call the *constant loss aversion* case:

$$v(x)=\alpha x \text{ if } x\le 0,\ v(x)=\beta x \text{ if } x\ge 0, \text{ with } \beta\ge 0 \text{ and } \alpha>\beta.$$

Under constant loss aversion as we define it, (3) and (4) apply with equality signs. We call *degree of loss aversion* the ratio β/α.

Note that v does not change over time (although the implied utility of wealth does).

2.3. CHOICE

We assume that the decision maker has free access to a loyal subordinate. Loyalty means that the subordinate is willing to do what he is asked to do. This a cooperative team situation (in the sense of Marschak and Radner, 1972), with no agency problem. *Centralization* occurs when the decision maker does not make use of the subordinate. *Decentralization* occurs when the subordinate is hired initially and the decision maker does not hear about the choice situation until the end of the second period.

Policies are chosen after a dynamic programming procedure that maximizes the sum of discounted expected utilities.

It may seem surprising that we use expected utility as a decision criterion rather than the full prospect theory approach (Tversky and Kahneman, 1992), that is usually associated with reference-dependent utility functions. Our primary reason for doing so is that we focus on those aspects of prospect theory which are not fundamentally incompatible with the von Neumann Morgenstern framework (see, e.g., Markowitz, 1952, and Fishburn and Kochenberger, 1979). Our purpose is to bring a dose of descriptive relevance in the classical normative model, and to discuss some implied prescriptions for dynamic choice.[6]

We first examine choice *under centralization*. The outcomes are perceived instantaneously when they are generated, and utility for x_1 (x_2) is experienced at the end

---

[5] $-X=\{-x{:}x\in X\}$ and $X\oplus X=\{x+y{:}x,y\in X\}$.

[6] Kahneman and Lovallo (1993) also use expected utility with reference-dependent utility functions, as an approximation of prospect theory, for the ease of exposition.

of the first (second) period. We assume throughout the chapter that perception immediately induces the corresponding utility.[7]

At the beginning of second period, expected utility under action $a_2(.|a_1, x_1)$ is thus

$$V_2(a_2(.|a_1, x_1)) = \sum_{x_2 \in X} v(x_2) a_2(x_2|a_1, x_1), \tag{6}$$

and the preferred action is

$$a_2^*(.|a_1, x_1) = \arg \max_{a_2 \in A_2(a_1, x_1)} V_2(a_2(.|a_1, x_1)). \tag{7}$$

Initially, anticipated satisfaction under action a_1 is therefore

$$V_1(a_1) = \sum_{x_1 \in X} \left[v(x_1) + \delta\, V_2(a_2^*(.|a_1, x_1)) \right] a_1(x_1), \tag{8}$$

where $\delta \in [0,1]$ is a discount factor whose interpretation is discussed below. The preferred first period action is

$$a_1^* = \arg \max_{a_1 \in A_1} V_1(a_1). \tag{9}$$

It is apparent that we assume time separability, i.e., a change in x_1 does not impact utility for any x_2 occurring in second period. This is a classical property in dynamic preferences modeling (see Meyer, 1976), and a tenet of dynamic programming.[8]

Under decentralization, the generated outcomes are disclosed together at the end of the second period (i.e., exposure to x_1 is delayed, in comparison with the centralization case). Therefore, expected utility associated with action $a_2(.\,|a_1, x_1)$ takes into account that the perceived outcome (at the end of period 2) will be the sum of realized x_1 and x_2.

We have

$$V_2^d(a_2(.|a_1, x_1)) = \sum_{x_2 \in X} v(x_1 + x_2) a_2(x_2|a_1, x_1), \tag{10}$$

and the preferred action is

$$a_2^{*d}(.|a_1, x_1) = \arg \max_{a_2 \in A_2(a_1, x_1)} V_2^d(a_2(.|a_1, x_1)). \tag{11}$$

Initially, the subordinate should pick

$$a_1^{*d} = \arg \max_{a_1 \in A_1} V_1^d(a_1), \tag{12}$$

given that the initially anticipated satisfaction under action a_1 is

$$V_1^d(a_1) = \sum_{x_1 \in X} \delta\, V_2^d(a_2^{*d}(.|a_1, x_1)) a_1(x_1). \tag{13}$$

A note on the interpretation of δ is in order. In our framework, the initial value of a given sequence (x_1, x_2) would be $v(x_1) + \delta v(x_2)$ under centralization, and $\delta v(x_1 + x_2)$ under decentralization. Now, a common rationale for discounting the future (in a context of wealth accumulation) is that an outcome received later has less time to

---

[7] One might argue that a decision maker should be able to exert willpower so as to postpone experience until the end of the second period, even when earlier perception has occurred. But our work precisely focuses on decision makers who don't have such willpower. Alternatively, our theory can be seen as an analysis of the conditions under which such willpower should prove valuable a priori.

[8] The descriptive relevance of time separability is discussed in Machina (1989), for instance.

fructify and should therefore be discounted. Why, then, do we apply a different discounting on x_1 under centralization than under decentralization? Indeed, x_1 is received anyway by the agent at the end of the first period (although it may not be perceived by the decision maker), and it can fructify identically under both centralization and decentralization. So, it may appear that our way of applying the discount factor is inconsistent.[9]

However, suppose that x_1 (assuming it is positive) is indeed invested. Such investment produces some additional payoff rx_1 at the end of the second period. Note that *this payoff should be seen as part of* x_2 if we conceive that $x_2 = rx_1 + x_2^{/}$, where $x_2^{/}$ is a second period outcome unrelated to the investment of previous gains. This is very much consistent with our notation, where second period lotteries are denoted by $a_2(\cdot|a_1, x_1)$. Now, recall that, under centralization, "the outcomes are perceived instantaneously when they are generated, and utility for x_1 (x_2) is experienced at the end of the first (second) period." Thus, x_1 has a direct impact through $v(x_1)$, plus an indirect impact at the end of the second period, through $v(rx_1 + x_2^{/})$. On the other hand, under decentralization, (x_1, $x_2^{/}$) will generate utility $v(x_1 + rx_1 + x_2^{/})$ at the end of the second period. Thus, in our model, δ is *not* meant to capture the aspect of time preferences that finds its rationale in the investment of early outcome.

But, then, what is the correct interpretation of δ? It is a subjective discount factor, a *psychological parameter* that reflects impatience towards the experience of positive utilities, and a preference for the postponement of negative utilities.[10] Since the decision maker's utility at each point in time is generated by exposure to the realized change in wealth, our evaluation functions are fully consistent with this interpretation.

When time preferences are motivated by investment activities only, then δ equals 1.

3. When to Decentralize

3.1. SUFFICIENT CONDITION FOR DECENTRALIZATION

The choice between centralization and decentralization is determined by the sign of $V_1^d(a_1^{*d}) - V_1(a_1^*)$. It is impossible to determine this sign without further specifying the dynamic choice problem. In the present section, we identify several classes of problems that guarantee a preference for decentralization.

[9] I am indebted to Bob Winkler who raised this issue.

[10] A psychological interpretation of discounting is not at all specific to the present work. See, e.g., the textbook by Malinvaud (1972, p. 234-35). A particular instance occurs in the theory of repeated games, where the discount factor is often interpreted as a probability of surviving the next period, i.e. the probability that the game will stop between any two consumption dates is 1-δ. Although values greater than 1 can be conceived, the traditional notion that decision makers are impatient is kept for the purpose of this chapter.

Our reasoning is as follows: if the best centralized policy brings more utility under decentralization, then decentralization dominates (i.e., the reference point should be stabilized).

To first illustrate the approach, let us examine the case of risk neutrality (i.e., $v(x) = kx$ for some $k > 0$). Note that there is no issue of stabilizing a reference point here, and decentralization only impacts utility through the discounting factor. Observe that, $\forall a_1 \in A_1$ and $x_1 : a_1(x_1) > 0$, we have $a_2^{*d}(.|a_1, x_1) = a_2^*(.|a_1, x_1)$.
Therefore, we can write

$$V_1(a_1^*) = \sum_{x_1 \in X}\left(kx_1 + \delta \sum_{x_2 \in X} kx_2\, a_2^*(x_2|a_1^*, x_1)\right)a_1^*(x_1), \tag{14}$$

and

$$V_1^d(a_1^*) = \sum_{x_1 \in X}\left(\delta \sum_{x_2 \in X} k(x_1 + x_2) a_2^*(x_2|a_1^*, x_1)\right)a_1^*(x_1)$$

$$= \sum_{x_1 \in X}\left(\delta\, kx_1 + \delta \sum_{x_2 \in X} kx_2\, a_2^*(x_2|a_1^*, x_1)\right)a_1^*(x_1). \tag{15}$$

Thus, $V_1^d(a_1^{*d}) - V_1(a_1^*) \geq V_1^d(a_1^*) - V_1(a_1^*) = (\delta - 1)\sum_{x_1 \in X} kx_1\, a_1^*(x_1)$. This justifies the following remark:

Remark. *Under risk neutrality, if the chosen first period action under centralization has negative expected outcome, then decentralization is preferred to centralization.*

Again, note that the advantage of decentralization here merely comes from postponing (expected) bad news.

Now, let us apply the same reasoning to the more realistic non-linear functions introduced in the previous section. Observe that

$$V_1^d(a_1^{*d}) - V_1(a_1^*) \geq V_1^d(a_1^*) - V_1(a_1^*)$$
$$\geq \sum_{x_1 \in X}\sum_{x_2 \in X}\left(\delta\, v(x_1 + x_2) - v(x_1) - \delta\, v(x_2)\right) a_2^*(x_2|a_1^*, x_1)\, a_1^*(x_1). \tag{16}$$

After defining the supports

$$X_1 = \{x_1 \in X : a_1^*(x_1) > 0\} \text{ and } X_2 = \{x_2 \in X : a_2^*(x_2|a_1^*, x_1) > 0, \text{ for some } x_1 \in X_1\}, \tag{17}$$

we can state:

Lemma 1. *If* $c(x_1, x_2) = \delta[v(x_1 + x_2) - v(x_2)] - v(x_1) \geq 0$ $\forall x_1 \in X_1$ *and* $\forall x_2 \in X_2$, *then decentralization is preferred to centralization.*

Consider the chosen plan of a centralized decision maker. Because of the shape of his utility function, the decision maker might prefer to perceive any possible first period outcome coupled with the second period outcome (rather than separately), knowing that the associated (dis)satisfaction will be postponed and discounted. When such is the case, decentralization of the centralized plan is beneficial, which is a sufficient (but not necessary) condition to favor decentralization.

3.2. UNDER CONSTANT LOSS AVERSION

For an individual with constant loss aversion, a loss is always more painful than the reduction of a gain by the same amount. Also, because there is no diminishing sensitivity, the utility associated with a loss remains unchanged when that loss is perceived on top of another loss. Moreover, postponement of losses is always welcome.

Proposition 1. *Under constant loss aversion, if the supremum of X_1 is non-positive, then decentralization is preferred to centralization.*

Proof. Consider any $x_1 \in X_1$ and $x_2 \in X_2$. Since $\sup X_1 \leq 0$, we have $x_1 \leq 0$. If $x_2 \leq 0$, then $c(x_1, x_2) = \delta\alpha(x_1 + x_2) - \delta\alpha x_2 - \alpha x_1 = (\delta - 1)\,\alpha x_1 \geq 0$. If $0 \leq x_2 \leq -x_1$, then $c(x_1, x_2) = \delta\alpha(x_1 + x_2) - \delta\beta x_2 - \alpha x_1 = (\delta - 1)\,\alpha x_1 + \delta(\alpha - \beta)x_2 \geq 0$. Else, $c(x_1, x_2) = \delta\beta(x_1 + x_2) - \delta\beta x_2 - \alpha x_1 = (\delta\beta - \alpha)x_1 \geq 0$. Thus, the condition in lemma 1 is always satisfied. □

Under constant loss aversion, a plain gain always provides less utility than the reduction of a loss by the same amount. Thus the decision maker can be better off by postponing exposure to a gain in order to reduce a loss in the future. However, postponing exposure to a gain for the sake of transforming it into the reduction of a loss in second period must be traded-off against the negative impact of gain's postponement, due to discounting. Moreover, when the second period loss is small in comparison with the postponed gain, the benefits of postponing exposure are limited. These trade-offs are analyzed by propositions 2 and 3.

Proposition 2. *Under constant loss aversion, if the supremum of X_1 is positive, $\delta \geq \frac{\beta}{\alpha}$, and $\sup X_2 \leq -\sup X_1$, then decentralization is preferred to centralization.*

Proof. Consider any $x_1 \in X_1$ and $x_2 \in X_2$. If $x_1 \leq 0$, then $c(x_1, x_2) \geq 0$, by the proof of proposition 1. If $x_1 > 0$ and $x_2 \leq -x_1$, then $c(x_1, x_2) = \delta\alpha(x_1 + x_2) - \delta\alpha x_2 - \beta x_1 = (\delta\alpha - \beta)\,x_1 \geq 0$ if $\delta \geq \frac{\beta}{\alpha}$. Thus the condition expressed in lemma 1 is always satisfied under the given conditions. □

Thus, when the discount factor is higher than the degree of loss aversion, and the second period possible outcomes under the best centralized plan are large losses, then decentralization dominates. When second period possible losses are not so large, we obtain a more restrictive condition on the discount factor.

Proposition 3. *Under constant loss aversion, if the supremum of* X_1 *is positive,* $\sup X_2 \in [-\sup X_1, 0]$, *and* $\delta \geq b(X_1, X_2) = \beta \sup X_1 / (\beta \sup X_1 + (\beta - \alpha) \sup X_2)$, *then decentralization is preferred to centralization.*

Proof. Consider any $x_1 \in X_1$ and $x_2 \in X_2$. If $x_1 \leq 0$, then $c(x_1, x_2) \geq 0$, by the proof of proposition 1. If $x_1 > 0$, $-x_1 \leq x_2 < 0$, and $\delta \geq b(X_1, X_2)$, then $c(x_1, x_2) = \delta\beta(x_1 + x_2) - \delta\alpha x_2 - \beta x_1 \geq 0$. If $x_1 > 0$, $x_2 < -x_1$, and $\delta \geq \frac{\beta}{\alpha}$, then we have $c(x_1, x_2) = \delta\alpha(x_1 + x_2) - \delta\alpha x_2 - \beta x_1 = (\delta\alpha - \beta)x_1 \geq 0$. Note that $b(X_1, X_2) > \frac{\beta}{\alpha}$ whenever $\sup X_2 \in [-\sup X_1, 0]$. □

3.3. UNDER DIMINISHING SENSITIVITY

When loss aversion is combined with diminishing sensitivity (i.e., when (3) and (4) apply with inequality signs), our claim that a loss is more painful than the reduction of a gain by the same amount must be qualified. Diminishing sensitivity implies that a large loss generates little dissatisfaction at the margin. Thus, little improvement comes from combining a loss with a small gain. Such little improvement might not compensate for the dropped satisfaction of perceiving the small gain in isolation. Therefore, combining a loss with a small gain is beneficial only when the loss is still sufficiently painful at the margin, i.e., when the loss is not too large. Note that, thanks to diminishing sensitivity, the utility associated with a loss always increases when that loss is cumulated with another loss.

Proposition 4. *Under loss aversion and diminishing sensitivity, if the supremum of* X_1 *is non-positive and* $\inf X_1 \geq -\inf X_2^+$, *then decentralization is preferred to centralization.*

Proof. Consider any $x_1 \in X_1$ and $x_2 \in X_2$. Since $\sup X_1 \leq 0$, we have $x_1 \leq 0$. If $x_2 \leq 0$, then, by (4), $c(x_1, x_2) \geq \delta v(x_1) - v(x_1) \geq 0$. If $x_2 \geq -x_1$, replacing $x_1 + x_2$ by $y \geq 0$, and given (3), $c(x_1, x_2) = \delta\left[v(y) - v(y + (-x_1))\right] - v(x_1) \geq \delta\left(-v(-x_1)\right) - v(x_1)$. Therefore, by (5) and given that $x_1 \leq 0$, we find $c(x_1, x_2) \geq 0$. Thus, when $\inf X_2^+ \geq -\inf X_1$, the condition expressed in lemma 1 is satisfied. □

When discounting is strong enough, the postponement of a first period loss is always attractive, according to proposition 5.

Proposition 5. *Under loss aversion and diminishing sensitivity, if the supremum of* X_1 *is non-positive and* $\delta \leq \min_{x_1 \in X_1} v(x_1) / \left[v(x_1) - v(-x_1)\right]$, *then decentralization is preferred to centralization.*

Proof. Consider any $x_1 \in X_1$ and $x_2 \in X_2$. Since sup $X_1 \le 0$, we have $x_1 \le 0$. From the proof of proposition 4, we know that both $x_2 \le 0$ and $x_2 \ge -x_1$ imply $c(x_1, x_2) \ge 0$. When $0 < x_2 < -x_1$, then $c(x_1, x_2) \ge 0$ if $\delta \le v(x_1)/[v(x_1 + x_2) - v(x_2)]$, by definition of $c(x_1, x_2)$. Note that $v(x_1)/[v(x_1 + x_2) - v(x_2)] > v(x_1)/[v(x_1) - v(-x_1)]$ $\forall x_2 \in]0, -x_1[$. Thus, if $\delta \le v(x_1)/[v(x_1) - v(-x_1)]$ $\forall x_1 \in X_1$, the condition in lemma 1 is satisfied. □

Diminishing sensitivity also imposes a qualification of proposition 2. Indeed, the reduction of a loss is not always more enjoyable than a plain gain by the same amount, especially if the loss is large and creates little dissatisfaction at the margin. Also, diminishing sensitivity implies that a gain yields less satisfaction when perceived on top of another gain than when perceived in isolation.

Proposition 6. *Under loss aversion and diminishing sensitivity, if the supremum of* X_1 *is positive,* $\delta = 1$ *and* $X_2 \subset [-\inf X_1^+, 0]$, *then decentralization is preferred to centralization.*

Proof. Consider any $x_1 \in X_1$ and $x_2 \in X_2 \subset [-\inf X_1^+, 0]$. For $x_1 \le 0$, we have that $c(x_1, x_2) \ge 0$, by proposition 4. If $x_1 > 0$ and $x_2 > 0$, then $c(x_1, x_2) < 0$, by (3). If $x_1 > 0$ and $x_2 = 0$, then $c(x_1, x_2) = 0$. If $x_1 > 0$ and $x_2 = -x_1$, then $c(x_1, x_2) = -v(x_1) - v(x_1) > 0$. When $-x_1 < x_2 < 0$, then replacing $x_1 + x_2$ by $y > 0$, and given (3) and (5), we obtain $c(x_1, x_2) = v(y) - v(x_2) - v(y + (-x_2)) \ge -v(-x_2) - v(x_2) \ge 0$. Thus, if $x_2 \in [-\inf X_1^+, 0]$ $\forall x_2 \in X_2$, then the condition expressed in the lemma is satisfied. □

4. When to Centralize

If the best decentralized policy brings more expected utility when centralized, then centralization dominates, i.e., shifts of the reference point should be allowed.

Observe that

$$V_1^d(a_1^{*d}) - V_1(a_1^*) \le V_1^d(a_1^{*d}) - V_1(a_1^{*d})$$
$$\le \sum_{x_1 \in X} \sum_{x_2 \in X} \left(\delta v(x_1 + x_2) - v(x_1) - \delta v(x_2)\right) a_2^{*d}(x_2 | a_1^{*d}, x_1) a_1^{*d}(x_1). \quad (18)$$

Now, define the supports

$$X_1^d = \{x_1 \in X : a_1^{*d}(x_1) > 0\},\ X_2^d = \{x_2 \in X : a_2^{*d}(x_2 | a_1^{*d}, x_1) > 0, \text{ for some } x_1 \in X_1^d\}. \quad (19)$$

We can state:

Lemma 2. *If* $c(x_1, x_2) = \delta[v(x_1 + x_2) - v(x_2)] - v(x_1) \le 0$ $\forall x_1 \in X_1^d$ *and* $\forall x_2 \in X_2^d$, *then centralization is preferred to decentralization.*

Decentralization postpones exposure to first period gains. Under discounting, this is considered a drawback.

Proposition 7. *Under constant loss aversion, if the infimum of* X_1^d *is non-negative and* $\delta \leq \frac{\beta}{\alpha}$, *then centralization is preferred to decentralization.*

Proof. Consider any $x_1 \in X_1^d$ and $x_2 \in X_2^d$. When $x_2 \geq 0$, $c(x_1, x_2) = \delta\beta(x_1 + x_2) - \delta\beta x_2 - \beta x_1 = (\delta - 1)\beta x_1 \leq 0$. If $x_2 < -x_1$, then $c(x_1, x_2) = \delta\alpha(x_1 + x_2) - \delta\alpha x_2 - \beta x_1 = (\delta\alpha - \beta)x_1$, which is negative if $\delta \leq \frac{\beta}{\alpha}$. If $-x_1 \leq x_2 < 0$, then $c(x_1, x_2) = \delta\beta(x_1 + x_2) - \delta\alpha x_2 - \beta x_1$ is negative if and only if $\delta \leq \beta x_1 / (\beta x_1 + (\beta - \alpha)x_2)$. As $\frac{\beta}{\alpha} \leq \beta x_1 / (\beta x_1 + (\beta - \alpha)x_2)$ when $-x_1 \leq x_2 < 0$, $\delta \leq \frac{\beta}{\alpha}$ is sufficient in this case too.□

Centralization also dominates under diminishing sensitivity, when the best decentralized policy leads to a sequence of gains.

Proposition 8. *Under loss aversion and diminishing sensitivity, if the infimum of both* X_1^d *and* X_2^d *is positive, then centralization is preferred to decentralization.*

Proof. By (3). □

Propositions 7 and 8 can trivially be used to generate necessary conditions for information decentralization.

5. Decentralization as an Anti-Conservatism Device

In section 2, we have characterized dynamic choice situations where a decision maker would benefit from information decentralization. Now, what can we say about the resulting (controlled) risk attitude? Is decentralized decision making more risk-taking when it arises? In this section, we introduce two classes of dynamic situations where the opportunity to decentralize information enhances risk taking.

5.1. ENTRY COST

Suppose that the decision maker faces a choice between the status quo (no loss, no gain over the two periods) and a risky action $a(.)$ that has positive expected payoff, but is only accessible after a first period sure loss ℓ. Such risky plan with entry cost could be a project in research and development, for instance.

Note that if (*i*) the decision maker has constant loss aversion or (*ii*) the decision maker is characterized by both loss aversion and diminishing sensitivity with

$$\text{either } \ell \geq -\inf\{x \in X^+ : a(x) > 0\} \text{ or } \delta \leq v(\ell)/v(\ell) - v(-\ell),\text{"}^{11} \tag{20}$$

then

$$\delta\left(v(\ell + x_2) - v(x_2)\right) - v(\ell) \geq 0 \quad \forall x_2 : a(x_2) > 0, \tag{21}$$

as shown in the proofs of propositions 1, 4 and 5. Thus, the risky plan yields higher anticipated expected utility under information decentralization than under centralization. Access to a hierarchical system renders choice of the risky plan more likely, because it softens the impact of the entry cost for a large class of decision makers.

5.2. REPEATED SITUATIONS

Now, assume that the decision maker must choose, at the beginning of each period, between the status quo and a given alternative *a*(.) that has a negative expected utility but a positive expected value. One can think of a decision whether to open a shop everyday in a stationary environment, for instance.

Solving the dynamic problem backward, we first observe that a centralized decision maker, at the beginning of the second period, would prefer the status quo over *a*(.). Initially, his choice thus reduces to *a*(.) played once versus the status quo, and he would stick with the status quo.

Under decentralization, the second period decision depends on the first period outcome. If *a*(.) has been picked initially, then the status quo is selected in second period if and only if the realized outcome x_1 was such that it is *a priori* beneficial to combine x_1 with *a*(.). In other words, status quo is selected if and only if

$$v(x_1) > \sum_{x_2 \in X} v(x_1 + x_2)a(x_2). \tag{22}$$

Thus, from the initial standpoint, information decentralization implies that any policy starting with *a*(.) has an anticipated value that is at least as great as the one associated with *a*(.) played once in second period. Indeed, by departing from the status quo initially, a decentralized decision maker anticipates

$$\sum_{x_1 \in X} \delta \max\left[v(x_1), \sum_{x_2 \in X} v(x_1 + x_2)a(x_2)\right] a(x_1). \tag{23}$$

If the above value is positive, then information decentralization leads the decision maker to take the same initial risk as a risk-neutral individual would take.

6. Escalation as a Side Effect

A possible side effect of information decentralization is escalation. Escalation is frequently observed in organizational behavior (Staw and Ross, 1989). It occurs when a

---

[11] i.e., the initial loss is not too large.

decision maker fails to terminate a project that proves unsuccessful. It generally results from a failure to consider sunk costs as really sunk. Because of diminishing sensitivity (which implies risk-seeking in the zone of losses) a decision maker "who has not made peace with his losses is likely to accept gambles that would be unacceptable for him otherwise" (Kahneman and Tversky, 1979, p. 287).

Under information decentralization, the decision maker commits not to get exposure to intermediary payoffs, and costs (losses) that are really sunk after the first period are not psychologically so. By stabilizing his reference point, the decision maker prevents himself to "make peace" with intermediary payoffs, and this may naturally lead to escalating policies.

Escalation and the underlying sunk cost effect are generally viewed as irrational. On the other hand, we have shown that information decentralization is often desirable and provides a mechanism to overcome loss aversion and resulting conservatism. Thus, we argue that escalation can often be seen as a side effect of an otherwise desirable and rational decision making process. The meta-decision about whether to postpone reports on intermediary outcomes often deals with a trade-off between a chance to avoid the status quo bias on the one hand, and a risk of escalation on the other hand.

7. Conclusion

This chapter has shown how, by allowing contingent policy implementation while maintaining the decision maker ignorant of the current decision being taken, a hierarchical device results in a transformation of individual risk preferences. This transformation can be anticipatively attractive for a decision maker.

We have not taken any normative position about the rationality of the features that we examined (loss aversion, reference point shifts, and the associated conservatism). We tend to respect them as real determinants of a decision maker's degree of satisfaction. However, as we have seen, the very target of improving utility may require a transformation of such features. In other words, the examined biases are sometimes self-defeating: a decision maker with reference point shifts may sometimes find it valuable to fight against such shifts. This aspect of our work is probably of particular interest to those economists who are concerned about whether individual biases survive on the market place (see the interesting discussion by Frey and Eichenberger, 1994). Overall, (and even though we only focused on a few of the most descriptively relevant features of human decision behavior) our work qualifies as prescriptive.

Our results also have an explanatory potential. We offered insights about why hierarchies exist and about how individual tendencies get transformed in the social context.

We found, for instance, that the possibility of initial losses (provided they cannot be too large) constitutes an incentive to switch from information centralization to information decentralization (proposition 4). Under constant loss aversion, information decentralization is always preferred, provided impatience is not too strong (propositions

1, 2 and 3). Moreover, access to a subordinate can help overcome conservatism in the presence of entry costs, as well as in repeated choice contexts. Escalation has been re-interpreted as a side effect of an otherwise desirable feedback system.

We have made rather specific assumptions regarding the utility of a sequence of outcomes: we assumed an additive multi-attribute utility function that favors first period utility. Other assumptions could be examined, but the basic contribution of this chapter would remain unchanged: when experienced utility depends on the frequency of exposure to the realized outcomes, a decision to decentralize information determines the chosen policies. Whether our set of assumptions is relevant is an empirical issue. Laboratory experiments should be conducted to see whether actual preferences correspond to the predictions made in section 3 to 5. At least, our general insight --that decentralization renders policies with possible intermediary losses more attractive-- corresponds well to Chandler's (1962) notion that decentralization is especially necessary when growth is not guaranteed.

Other assumptions about the nature and dynamics of reference points might also be examined. We assumed that the reference point is the current known wealth position and that the decision maker gets immediately accustomed to his/her new wealth level, when it is disclosed. Reference points can also be expectation/aspiration levels, and reference point updating could be slower than assumed here. The impact of information decentralization on preferences under these assumptions remains an open question.

Future research might also relax the assumption of a free and loyal subordinate. Indeed, the decision maker's preferences over the timing of communication in the hierarchy might conflict with the best interest of the subordinate, for some reason.

References

Arkes, H.R. (1991) Costs and benefits of judgment errors: implications for debiasing, *Psychological Bulletin* **110**, 486-498.

Barnard, Ch. (1968) *The Functions of the Executive,* Harvard University Press, Cambridge.

Bell, D.E. (1985) Disappointment in decision making under uncertainty, *Operations Research* **30**, 961-981.

Chandler, A.D., Jr. (1962) *Strategy and Structure,* MIT Press, Cambridge.

Elster, J. (1979) *Ulysses and The Sirens: Studies in Rationality and Irrationality,* Cambridge University Press, Cambridge.

Fishburn, P.C. and Kochenberger, G.A. (1979) Two-piece Von Neumann-Morgenstern utility functions, *Decision Sciences* **10**, 503-518.

Fishburn, P.C. (1988) *Nonlinear Preference and Utility Theory,* Johns Hopkins University Press, Baltimore.

Frey, B.S. and Eichenberger, R. (1994) Economic incentives transform psychological anomalies, *Journal of Economic Behavior and Organization* **23**, 215-234.

Geanakoplos, J. and Milgrom, P. (1991) A theory of hierarchies based on limited managerial attention, *Journal of the Japanese and International Economies* **5**, 205-225.

Kahneman, D. and Tversky, A. (1979) Prospect theory: an analysis of decision under risk, *Econometrica* **47**, 263-291.

Kahneman, D. and Lovallo, D. (1993) Timid choices and bold forecasts: a cognitive perspective on risk taking, *Management Science* **39**, 17-31.

Kreps, D.M. and Porteus, E.L. (1978) Temporal resolution of uncertainty and dynamic choice theory, *Econometrica* **46**, 185-200.

Loewenstein, G.F. (1988) Frames of mind in intertemporal choice, *Management Science* **34**, 200-214.

Machina, M.J. (1989) Dynamic consistency and non-expected utility models of choice under uncertainty, *Journal of Economic Literature* **27**, 1622-1668.

Malinvaud, E. (1972) *Lectures on Microeconomic Theory*, North-Holland, Amsterdam.

Markowitz, H. (1952) The utility of wealth, *Journal of Political Economy* **60**, 151-158.

Marschak, J. and Radner, R. (1972) *Economic Theory of Teams*, Yale University Press, New Haven.

McClennen, E.F. (1990) *Rationality and Dynamic Choice: Fundational Explorations*, Cambridge University Press, Cambridge.

McCrimmon, K.R. and Wehrung, D.A. (1986) *Taking Risks*, The Free Press, New York.

Meyer, R.F. (1976) Preferences over time, in R.L. Keeney and H. Raiffa, *Decisions with Multiple Objectives: Preferences and Value Tradeoffs*, John Wiley, New York.

Postrel, S. and Rumelt, R.P. (1992) Incentives, routines, and self-command, *Industrial and Corporate Change* **1**, 397-425.

Radner, R. (1992) Hierarchy: the economics of managing, *Journal of Economic Literature* **30**, 1382-1415.

Samuelson, W. and Zeckhauser, R. (1988) Status quo bias in decision making, *Journal of Risk and Uncertainty* **1**, 7-59.

Schelling, T.C. (1978) Egonomics, or the art of self-management, *American Economic Review* **68**, 290-294.

Staw, B.M. and Ross, J. (1989) Understanding behavior in escalation situations, *Science* **246**, 216-220.

Stigler, G. and Becker, G. (1977) De gustibus non est disputandum, *American Economic Review* **67**, 76-90.

Strotz, R.H. (1956) Myopia and inconsistency in dynamic utility maximization, *Review of Economic Studies* **23**, 165-180.

Thaler, R.H. (1985) Mental accounting and consumer choice, *Marketing Science* **4**, 199-214.

Tversky, A. and Kahneman, D. (1991) Loss aversion in riskless choice: a reference-dependent model, *Quarterly Journal of Economics* **106**, 1039-1061.

Tversky, A. and Kahneman, D. (1992) Advances in prospect theory: cumulative representation of uncertainty, *Journal of Risk and Uncertainty* **5**, 297-323.

ALLAIS PHENOMENA AND COMPLETENESS OF PREFERENCES

S. Ríos*, S. Ríos-Insua, D. Ríos Insua, J. Pachón**,
M.A. Pacios+, P. García Barreno*
*Real Academia de Ciencias
**Decision Analysis Group, Universidad Politécnica de Madrid
+Departamento de Economía, Universidad Carlos III de Madrid

Abstract

Most descriptive failures of Expected Utility Theory have been placed on the independence postulate. For a while, there was some debate on whether the blame should be put on the completeness postulate. We reenter into that debate providing experimental support for a robust theory of decision making. Consequences for decision analysis practice are outlined.

KEYWORDS: Expected Utility, Allais Phenomena, Independence, Completeness, Classes of Utility Functions.

1. Introduction

It is now more than fifty years since the publication of the seminal work of Von Neumann and Morgenstern (1944). Expected Utility (EU) Theory remains as a leading paradigm for Decision Making under risk, in spite of many criticisms as an invalid descriptive theory, starting from Allais paradox. These criticisms, in turn, have led to many new theories which attempt to improve upon EU from a descriptive point of view, concentrating mainly on relaxations of the independence principle. Recent reviews include Weber and Camerer (1987) and Rios Insua (1994). Whereas experience concerning the descriptive inadequacy of EU has accumulated, recent comparisons among new theories have painted a somewhat different picture.

Note first that by abandoning axioms for descriptive purposes, we have to abandon them for normative purposes: Some theories have been proposed a bit too lightly, since they violate principles like transitivity or stochastic dominance. For example, Sarin (1992) suggests that alternative theories may run into problems in dynamic decision contexts, since their solutions for extensive and normal forms may not coincide.

Moreover, new experiments have been carried out: for example, Camerer (1992) observes that all alternative theories so far proposed run into one kind or another of descriptive problems; Currim and Sarin (1989), comparing Prospect Theory

R. Nau et al. (eds.), Economic and Environmental Risk and Uncertainty, 245–256.

(PT) and EU, conclude that whereas PT is predictively better for paradoxical situations, both perform equally well on the whole; Bernasconi (1992) observes that performers of experiments seem to conform progressively to EU postulates, after repetitions of experiments; finally, Hey and Orme (1994) observe that, for many subjects, EU fits equally well as other theories from a statistical point of view, whereas economic implications are not that important for those for which fit is not so good. To sum up, risky decisions do not fully conform to EU, but they approximate it very reasonably.

Some recent descriptive theories are close to this idea. For example, Leland (1994) suggests that paradoxes may not be due to the preference structure but rather to cognitive constraints and experiential limitations, suggesting a resolution of EU violations based on similarity judgements. A related issue is that of imprecision in judgements, a current favourite topic in Bayesian Analysis, known as Robust Bayesian Analysis, which generalises statistical and decision models to the case in which there is a class of prior distributions (rather than one) and/or a class of utility functions (rather than one). Actually, some of the philosophers of this current, see e.g. Seidenfeld (1988), have discussed the pros and cons of dropping the independence and the completeness principles, with a view towards dealing with descriptive failures of EU.

Our work fits into this last tradition and stems from a previous study reported in Rios et al (1994). We have conducted experiments aimed at explaining paradoxical behaviour in Allais type experiments via incomplete preferences modelled through a class of utility functions: Most responses, including paradoxical ones, may be explained in such a way. Moreover, those responses which we are not able to account for refer to pairs of alternatives with small difference in expected utility. Our results contrast, therefore, with conclusions in related studies.

The paper is structured as follows: we first provide a description of foundational issues concerning a robust theory of Decision Making and how it might cope with paradoxes. We then describe the procedures and phases involved in our experiments. Section 4 presents general results followed by explanations of paradoxes via classes of utility functions. We conclude with some reflections for the practice of Decision Analysis and discussions.

2. Robust Decision Theory and Allais Paradox

The Allais paradox and other accumulated experimental evidence suggest that EU theory is not valid descriptively. Many authors view this as a challenge to the principle of independence. Following our suggestion in Rios et al. (1994), we shall analyse whether Allais type paradoxes may be in part attributed to incomplete preferences.

We shall first review briefly the foundations of a theory of Decision Making under risk allowing for incompleteness in preferences and then show how it may cope with some paradoxes.

2.1. FOUNDATIONS FOR ROBUST DECISION MAKING

Most generalisations of EU theory depart from Von Neumann and Morgenstern axioms by modifying and/or dropping one of their postulates. Specifically, most work trying to account for Allais phenomena modify and/or drop the independence principle.

When performing Decision Analysis, it is often the case that at least in the first stages of analysis the DM may find some alternatives incomparable. Similarly, when assessing a utility function, he may find it difficult to provide a precise value for the utility of a consequence: He may be more comfortable providing some range for that value. We are thus interested in cases in which the DM's preferences are incomplete (or imprecise). This leads us to drop the completeness principle from the Von Neumann-Morgenstern postulates. By modifying their continuity condition we may provide Theorem 1, taken from Rios Insua and Martin (1994), which models the case we are interested in.

Theorem 1 *Let $\mathcal{C} \subset \mathbf{R}^n$, $\mathcal{F}$ be a convex set of distribution functions whose support is compact and contained in $\mathcal{C}$, and includes the set of degenerate distribution functions. Let $\preceq$ be a binary relation in $\mathcal{F}$. Then*

U1. $(\mathcal{F}, \preceq)$ *is a quasi order.*

U2. For $\alpha \in (0,1), F, G, H \in \mathcal{F},\ F \preceq G \iff \alpha F + (1-\alpha)H \preceq \alpha G + (1-\alpha)H$.

U3. For $F, G, H, L \in \mathcal{F},\ (\alpha F + (1-\alpha)G \preceq \alpha H + (1-\alpha)L, \forall \alpha \in (0,1]) \Longrightarrow G \preceq L$.

are equivalent to the existence of a class $\mathcal{U}$ of real functions u on $\mathcal{C}$ such that

$$F \preceq G \iff \left(\int u dF \leq \int u dG, \forall u \in \mathcal{U} \right). \tag{1}$$

Thus, we may model incomplete preferences under risk by means of a class of expected utilities based on a class of utility functions. Conventional methods of utility assessment, such as the *probability equivalent* (PE) method, see Farquhar (1984), may be modified to assess the corresponding class of utility functions: rather than demanding precise PE's, we aim only at getting upper and lower bounds for them. For example, the PE method requires the DM to provide a probability p such that receiving a best consequence c^* with probability p and a worst consequence c_* with probability $(1-p)$ is indifferent to receiving a consequence c with certainty. Rather than asking him to state a precise p, we could ask him to state a range $[p', p'']$ such that if $p < p'$ $(p'' < p)$ he definitely prefers the certain consequence (the lottery), whereas if $p' < p < p''$ he may not be so sure of his preferences. Von Nitzsch and Weber (1988) and Kimbrough and Weber (1994) describe programs for utility assessment, including the assessment of classes of utility functions. We describe our own program below.

2.2. ROBUSTNESS AND ALLAIS PARADOX

Our connection between Allais paradox and incomplete preferences stems from the observation that if subjects do not respond to utility assessment questions in a way that leads to a precisely specified utility function, then we should not be surprised if their responses to choice questions are sometimes inconsistent with the existence of well-defined utilities. In such cases, the subjects presumably must rely on heuristic decision rules of some kind possibly with different strategies for choosing in different cases, as suggested by Leland (1994). See also comments in Bernasconi (1992) in that direction. Thus, we shall use two descriptive models to try to explain Allais paradox via incomplete preferences, as we illustrate with Allais classical example, which we recall briefly.

We offer a subject a choice between lotteries $A = (1, 1M)$, where $1M$ stands for 1 million pesetas, and $B = (.01, 0M; .89, 1M; .10, 5M)$. We then offer him a choice between lotteries $C = (.89, 0M; .11, 1M)$ and $D = (.90, 0M; .10, 5M)$. Typically, see MacCrimonn and Larsson (1979), a nonnegligible proportion of respondents choose A in the first case and D in the second case. Assuming an EU maximizer and $u(0M) = 0$, $u(5M) = 1$, preference of A over B in the first problem leads to the inequality $u(1M) > .1/.11$, whereas preference of D over C in the second one leads to $u(1M) < .1/.11$. Since both inequalities are incompatible, EU theory is unable to describe those choices.

2.2.1. *Descriptive model I*

Suppose the DM has incomplete preferences, modelled by $u \in \mathcal{U}$, and are such that $u(1M) \in [a, b]$. We may still assume that $u(0) = 0$, $u(5M) = 1$, $\forall u \in \mathcal{U}$. The expected utility of A, say $v(A)$, satisfies $v(A) \in [a, b]$. Similarly, $v(B) \in [.89a + .1, .89b + .1]$, $v(C) \in [.11a, .11b]$, and $v(D) = .1$. The difference in expected utility between A and B is $v(A) - v(B) \in [.11a - .1, .11b - .1]$; $v(C) - v(D)$ lies in the same interval. Suppose now that $.1/.11 \in [a, b]$. Then, $0 \in [.11a - .1, .11b - .1]$ and we could argue that such imprecision might describe the inconsistency: in his first choice of A over B, he would be answering in accord with a $u \in \mathcal{U}$ such that $u(1M) > .1/.11$; in his second choice, with a $u \in \mathcal{U}$ such that $u(1M) < .1/.11$.

Note that $u \in \mathcal{U}$ identifies cognitive constraints of the DM; their assessment is a task which demands significant cognitive effort on behalf of the DM. On the contrary, he typically faces an Allais type experiment in an unaided fashion, so he acts as if he picked, via some heuristic or randomly, different utilities from the set $\mathcal{U}$ representing his preferences. Admittedly, some people might say that this is just a restatement of the problem. However, note we may counter it if we show that for consistent DM's $.1/.11 \notin [a, b]$. Specifically, suppose we assess $.1/.11 < a$. Then, we could predict that he would prefer A to B and C to D, and, viceversa, if $.1/.11 > b$.

To wit, consider the example of one of the participants in our experiments. Table 1 includes the assessment of his (imprecise) preferences where the attribute is pesetas, within two different parts of the attribute space. Our assessment was based in the modification of the PE method described above:

TABLE 1: Utility assessments for a participant. Consequences in pesetas

Pesetas	0	100	500	10000
Utility	0	[.10,.20]	[.30,.35]	1
Pesetas	0	$1M$	$2M$	$5M$
Utility	0	[.75,.80]	[.87,.92]	1

For example, with $u(0) = 0$, $u(5M) = 1$, we conclude that $u(1M) \in [.75, .8]$.

Let us consider now his answers to a number of Allais type problems, with probabilities as in the classical example and consequences (e, f, g), with e, worst; f, intermediate; and g, best consequence, respectively. When $(e, f, g) = (0, 100, 500)$, he answered consistently leading to inequality $.1u(500) > .11u(100)$, which holds always for any assessment consistent with the bounds in the table. However, he provides inconsistent answers when $(e, f, g) = (0, 2M, 5M)$ leading to inequalities $.1u(5M) > .11u(2M)$, and $.1u(5M) < .11u(2M)$, respectively. There are assessments consistent with both inequalities, given the bounds in the tables, since $.1/.11 \in [.87, .92]$.

2.2.2. *Descriptive model II*

For our second descriptive model, we assume again that the DM's preferences are modelled by $u \in \mathcal{U}$, with $u(1M) \in [a, b]$, $u(0M) = 0$, $u(5M) = 1$. The typical violation of EU leads to inequalities

$$u(1M) > .89u(1M) + .10,$$

$$.11u(1M) < .10.$$

Suppose now that rather than comparing directly both lotteries, the DM does it in an indirect way, thinking first of the certainty equivalents of lotteries and, then, comparing both CE's. To account for the violation, suppose that, by some heuristic or random mechanism, the DM draws a value $x \in [a, b]$ and a value $y \in [.89 \times a + .1, .89 \times b + .1]$, and compares them to find that $x > y$, replying A preferred to B. In the second case, he draws a value $z \in [.11 \times a, .11 \times b]$ and compares it with .10, finding that $z < .10$ and replying D preferred to C.

Consider one of the performers in one of our experiments, who replied with the typical inconsistent pattern and $u(1M) \in [.75, .78]$. For the second question, since $.11 \times .78 < .1$, we predict that the performer will reply as he does. For the first case, the performer draws from $[.75, .78]$ and then from $[.7675, .7942]$. Clearly, there are drawings compatible with the choice of A over B.

2.2.3. *Comments*

Our axioms in Theorem 1 lead to ranking alternatives in a Pareto sense, as described by (1), that is, G is preferred to F if and only if the expected utility of G is bigger than the expected utility of F, for each utility function u in the class

$\mathcal{U}$. Therefore, in the Allais type cases we are considering, those axioms would suggest incomparability of alternatives. Their role in our descriptive models is to identify the constraints defining the class of utility functions modelling the DM's preferences. They may be used also to identify potential inconsistencies, much as we shall use them below. The descriptive models suggested come into play since performers have to answer to experiments without much reflection and with little decision aid, being therefore guided by psychological concerns and emotional effects, in which other descriptive models, such as those in Leland (1994) might have their role. Sarin (1992) has some comments in this direction.

Note that both models correspond to different choice strategies. The first case corresponds to one in which the DM compares both alternatives and, therefore, uses a same utility function, chosen from $u \in \mathcal{U}$, for both alternatives. We may call this model *comparative* or *fixed utility model.* In the second case, the DM evaluates first the alternatives and then compares the evaluations, therefore not using the same utility for both evaluations, but still within the set $u \in \mathcal{U}$. We may call this model *evaluative* or *variable utility model.* Obviously, the second one is more permissive than the first one, in that it allows more violations of EU.

3. Our Experiments

We first conducted an initial experiment with one hundred and thirty-one undergraduate students, from the School of Computer Science of Madrid Technical University. They had ample quantitative training, knowledge of Probability, but not of Decision Analysis. The results of that experience have been reported in Rios et al (1994). That study allowed us to acquire experience with the techniques and models employed, and control some factors affecting the experiments, like the order of questions and presentation form.

3.1 EXPERIMENTS WITH ARMY OFFICERS

Our first experiment was conducted with a group of twenty-three young officers of the Spanish Army. All of them had undergraduate degrees in quantitative fields (Maths, Physics, Engineering). First, they passed a block of Allais type problems in a booklet including:

- Five problems with monetary consequences (pesetas), three of them with small (0,100,500) to moderate (0,50000,250000) gains, two with large gains ((0,1M,5M) and (0,2M,5M)). Probabilities were as in Allais' classical example.

- Five problems of military type, specifically choosing between commanding different UN peacekeeping missions attaining their military objectives, but with chances of having several deaths. Probabilities were as in Allais' classical example.

Problems were presented in randomised order. Subjects were given as much time as required and verbal and written instructions were provided. After the tests, they passed an eight hour course in Decision Analysis, which included an exposure to paradoxes in rationality. Then, they passed again the tests. Finally, we assessed their utility functions both for monetary and military problems. For that purpose, a program was prepared based on the modification of the PE method described in Section 2.1, in order to get a class of utility functions. The basic idea was to bound interactively above and below the PE's of the performers until a range was determined in which the PE should lie, but such that they were not able to further specify it.

Participants volunteered for the experiment, which provided several stimuli like receiving training in a subject new to them (Decision Analysis), receiving credits for their career and changing from their habitual routines.

A larger sample of seventy-nine additional officers passed only the tests.

3.2. EXPERIMENTS WITH PHYSICIANS

A similar experiment was conducted with twenty-two physicians from the University Hospital Gregorio Marañón. Their quantitative training was more limited. Again they went through a session of tests/course/tests/utility assessment, but this last phase only for twenty of them. They passed the same tests with monetary consequences. They also passed six tests with medical consequences, specifically, referring to choice between treatments against flu epidemics, with average number of working days lost as consequence. Again they volunteered for the study, with similar stimuli as the officers.

3.3. EXPERIMENTS WITH LECTURERS

A final experiment took place with fifty-two lecturers from the Department of Economics of Universidad Carlos III, who volunteered for the study. They only passed the monetary tests and twenty-two of them had their classes of utility functions assessed, with the aid of our program.

4. General Results

Table 2 provides the proportions of consistent performers for monetary tests for physicians, before (Meds1) and after (Meds2) the course; the small group of officers, before (Mils1) and after (Mils2) the course; the whole group of officers; the group of lecturers and, for completeness, the group of students. The first row includes the intermediate prize and the best prize in each test, the worst prize being 0 pesetas. For example, within the group of physicians after the DA course (Meds2), when the intermediate prize was 50000 pesetas and the best prize 250000 pesetas in Allais paradox, 86.9% of performers answered consistently with EU.

TABLE 2: Proportion of consistent respondents. Monetary consequences

Prizes	100,500	10^4,5×10^4	5×10^4,25×10^4	1M,5M	2M,5M	Size
Meds1	86.9	78.2	70.0	86.9	78.2	22
Meds2	91.3	95.6	86.9	91.3	95.6	22
Mils1	78.3	74.6	78.3	83.2	74.6	23
Mils2	78.3	87.2	78.3	74.2	83.2	23
Mils	76.5	74.5	70.5	77.5	70.5	101
Lects	86.5	80.7	80.7	75.0	69.2	52
Studs.	77.8	74.8	76.3	64.1	58.7	131

We analysed proportions via classical equality of proportions test, either via normal approximations, when appropriate conditions hold, or alternatively, via Fisher exact tests. Though not central in our research, we discuss some of the main findings, since they are relevant in our search. First, note that there is a deterioration in student performance from small prizes (76.3%) to large prizes (64.1%). As the performers mentioned, this is probably due to finding the questions more difficult, since prizes in millions of pesetas refer to amounts of money they are not used to deal with. On the contrary, we have perceived no such deterioration in performance in the rest of participants, perhaps due to their being more senior and, consequently, more used to larger amounts of money and having more precise preferences in that range of gains.

There is also a significant improvement of performance after the DA course for physicians. There seems to be an improvement in performance for officers, but it is not significant, perhaps because they find more difficult to change their opinions. All in all, performance were not too disparate from similar classical studies, like MacCrimmon and Larsson (1979), which indicated that our stimuli where good enough.

A similar picture holds for nonmonetary questions. Table 3 provides proportions of consistent performers in medical and military questions, before and after the DA course.

TABLE 3: Proportion of consistent respondents. Nonmonetary consequences

Question	Q1	Q2	Q3	Q4	Q5	Q6
Meds1	86.9	78.2	69.5	91.3	78.2	91.3
Meds2	86.9	86.9	86.9	86.9	82.6	91.3

Question	Q1	Q2	Q3	Q4	Q5
Mils1	78.2	65.2	60.8	60.8	56.5
Mils2	82.6	65.2	69.5	65.2	52.1

In both cases, we observe some improvement in performance, though not significant in general. Also performance is worse for officers in military questions than in monetary questions. When asked about that, they mentioned the bigger complexity of those questions, in which consequences are numbers of deaths in a mission. The opposite happens for doctors in medical questions, who found these easier than the monetary ones.

5. Explanation via Utility Functions

The final part of the experiments included the assessment of utility functions for twenty physicians, twenty-three officers and twenty-two lecturers. Rather than trying to assess a precise utility function, we assessed a class of utility functions with the aid of the program described above. Our aim with this phase was to check whether inconsistencies in Allais type problems may be attributed to incomplete preferences as suggested in the descriptive models above.

Table 4 summarises the results of this part of experiment. It includes results for physicians both for monetary and medical problems; for officers, both for monetary and military problems, and for lecturers. Of all the decision problems posed with valid answers (No. Que.), we analysed how many answers may be explained via our first descriptive model (DI) or fixed utility model; how many answers cannot be explained with the first one but can be explained by the second one (DII) or variable utility model; how many responses are not predicted by our models but are such that the difference in expected utility between incumbent alternatives is smaller than 0.01 ($\leq .01$); how many are such that the difference in expected utility is bigger than 0.01 ($> .01$) and the biggest difference in expected utility (Max).

TABLE 4: Explanation of answers

	No.Que.	DI	DII	$\leq .01$	$> .01$	Max
Med.Mon.	198	148	42	6	2	.08
Med.Med.	240	208	20	6	6	.08
Mil.Mon.	230	200	11	16	3	.05
Mil.Mil.	230	195	21	13	1	.03
Lecturers	220	173	33	9	5	.03

We believe that the table draws a very suggestive picture. Note first that both the Descriptive Model I (75%, 85%, 87%, 85%, 77%) and, specially, Descriptive Model II (95%, 94%, 92%, 94%, 93%) allows a significant proportion of responses in the questions posed. It is specially interesting to analyse inconsistent answers, since all consistent answers are predicted by our models, though in some cases we have to use Descriptive Model II. Table 5 summarises results.

TABLE 5: Inconsistent answers in Allais type problems

	No.Inc.	Predicted
Med.Mon.	9	5
Med.Med.	15	9
Mil.Mon.	23	12
Mil.Mil.	40	29
Lecturers	25	18

For example, officers in military problems provide 40 inconsistent answers in Allais type problems. Of those, 29 may be described by our models, typically by the second one.

Thus, to some extent, we might claim that incomplete preferences have some role in explaining inconsistencies in Allais type problems. What we, therefore, identify are possible sources of inconsistency.

Note also that those responses not consistent with our models correspond to cases in which the difference in expected utility between alternatives is very small. Typically, in this situation we should perform sensitivity analysis and, possibly, upon reflection the DM might find little difference between both alternatives. See Leland (1994) for further discussion concerning comparison of alternatives which are close in a expected utility sense.

6. Conclusions and Discussion

Stemming from recent work in Robust Bayesian Analysis, we have suggested that incomplete preferences might have a role in explaining Allais phenomena. Through a careful and aided process of utility assessment, we are only able to get a class of utility functions modelling the DM's incomplete preferences. The assessment is based in comparing pairs of simple alternatives. On the other hand, in Allais type experiments we typically offer the DM pairs of more complex alternatives, very frequently of similar expected utility, and he has to choose in an unaided fashion between them. Therefore, inconsistencies should not be at all surprising.

Our descriptive models seem to be able to cope with many responses in tests. Descriptive Model I is closer to the model suggested in Theorem 1, in that it is based in differences of expected utility with respect to a same utility function $u \in \mathcal{U}$. The range of expected utility differences may suggest possible inconsistencies in Allais type problems, which may be explained in terms of other descriptive models. Descriptive Model II is motivated by conventional Decision Analysis practice: once we have assessed a utility function, we proceed by computing expected utilities of alternatives and choosing the one with maximum EU. In our case, we have a class $\mathcal{U}$ of utility functions, and we proceed by computing the range of expected utilities for each alternative. These may suggest possible inconsistencies in Allais type problems, which again may be explained in terms of other descriptive models.

Of all descriptive models, those in Hey and Orme (1994) are possibly the closest to our suggestions. Basically, what we do by assessing the class $u \in \mathcal{U}$ is to identify the support (or a subset of the support with high probability) of the distributions involved in their models. Note, however, that they propose normal distributions. Given our normalisation, we would have to go for distributions with compact support. It would be perhaps interesting to continue our research in that direction.

One could argue that typical Allais violations are not random, in that some choice pattern typically emerges. Kadane (1992) suggests an explanation for Allais phenomena based on Bayesian processing of the probabilities in the examples. In that sense, Allais violations would not be random, since once probabilities have been processed, choice is in accord with EU. Note though that such inconsistent pattern does not always emerge, giving some room for random choice theories.

Allais type problems constitute an extremely useful source of problems given our conception of Decision Analysis. Once we have structured the problem, we go through preference and belief assessment through classes of utility functions and probability distributions. Sensitivity analyses allow us to detect possible inconsistencies (e.g., by posing Allais type problems based on the classes of utility functions detected) and disparities (e.g., by checking whether different judgements lead to very different optimal alternatives) in the DM's judgements. If this is the case, we should go for further refinement of the classes, and proceed again through new cycles of Decision Analysis until the DM feels the model is *requisite* for making a decision (Phillips, 1984).

Acknowledgments Work supported by a CICYT project and a BBV Foundation project. We are grateful to discussions with R. Nau, F. Cañizo and M. Illán and the computational work of V. González.

References

Bernasconi, M. (1992) Different frames for the independence axiom: an experimental investigation in individual decision making under risk, *Journal of Risk and Uncertainty*, 5, 159-174.

Camerer, C. (1992) Recent tests of generalizations of Expected Utility Theory, in Edwards (ed) *Utility Theories: Measurement and Applications*, Kluwer

Currim, I. and Sarin, R. (1989) Prospect versus utility, *Mgt. Science*, 35, 22-41.

Farquhar, P. (1984) Utility assessment methods, *Mgt. Science*, 30, 1283-1300.

Hey, J. and Orme, C. (1994) Investigating generalizations of expected utility theory using experimental data, *Econometrica*, 62, 1291-1326.

Kadane, J. (1992) Healthy skepticism as an expected utility explanation of the phenomena of Allais and Ellsberg, *Theory and Decision*, 32, 57-64.

Kimbrough, S. and Weber, M. (1994) An empirical comparison of utility assessment programs, *to appear in EJOR.*

Leland, J. (1994) Generalized similarity judgements: an alternative explanation for choice anomalies, *Jour. Risk and Uncertainty*, 9, 151-172.

MacCrimmon, K. and Larsson, S. (1979) Utility theory axioms or paradoxes, in Allais and Hagen (eds) *Expected Utility Theory and the Allais Paradox*, Reidel.

Phillips, L. (1984) A theory of requisite decision models, *Acta Psychologica*, 56, 29-48.

Rios, S., Rios-Insua, S., Rios Insua, D. and Pachon, J. (1994) Experiments in robust decision making, in Rios, Rios Insua, Rios-Insua (eds) *Decision Theory and Decision Analysis: Trends and Challenges*, Kluwer.

Rios Insua, D. (1994) Ambiguity, imprecision and sensitivity in Decision Theory, in Puri and Vilaplana (eds) *New Progress in Probability and Statistics*, SVP.

Rios Insua, D. and Martin, J. (1994) On the foundations of robust decision making, in Rios, Rios Insua, Rios-Insua (eds) *Decision Theory and Decision Analysis: Trends and Challenges*, Kluwer.

Sarin, R. (1992) What now for generalized utility theory, in Edwards (ed) *Utility Theories: Measurement and Applications*, Kluwer.

Seidenfeld, T. (1988) Utility theory without independence or without ordering, *Economics and Philosophy*, 4, 267-315.

Von Neumann, J. and Morgenstern, O. (1944) *Theory of Games and Economic Behavior*, Princeton UP.

Von Nitzsch, R. and Weber, M. (1988) Utility function assessment on a micro-computer: a reliable, interactive procedure, *Annals of Operations Research*, 16, 149-160.

Weber, M. and Camerer, C. (1987) Recent developments in modelling preferences under risk, *OR Spectrum*, 9, 129-151.

REPRESENTING PREFERENCE RELATIONS WITH NONTRANSITIVE INDIFFERENCE BY A SINGLE REAL-VALUED FUNCTION

GIANNI BOSI and ROMANO ISLER
Dipartimento di Matematica applicata "Bruno de Finetti"
Piazzale Europa 1, Trieste, 34127, Italy

Abstract

Let $\prec$ be an interval order on a topological space (X, τ), and let $x \prec^* y$ $(x \prec^{**} y)$ if and only if there exists ξ (there exists η) in X with $x \prec \xi \precsim y$ (respectively, $x \precsim \eta \prec y$). Then $\precsim^*$ and $\precsim^{**}$ are complete preorders. In the particular case when $\prec$ is a semiorder, let $x \prec^0 y$ if and only if either $x \prec^* y$ or $x \prec^{**} y$. Then $\precsim^0$ is a complete preorder, too. We present sufficient conditions for the existence of continuous utility functions representing $\precsim^*$, $\precsim^{**}$ and $\precsim^0$, by using the notion of strong separability of a preference relation, which was introduced by Chateauneuf.

1. Introduction

The problem of representing numerically preference relations with nontransitive indifference on a topological space (X, τ) was extensively treated in literature. Chateauneuf (1987) found out necessary and sufficient conditions for the existence of a pair of continuous functions u, v representing an interval order $\prec$ on a connected topological space (X, τ) (in the sense that $x \prec y$ if and only if $u(x) < v(y)$). Bridges (1985) showed that an interval order $\prec$ may be represented by a single real-valued function f. If $\prec$ is an interval order and we let $[x \precsim^* y$ if and only if $(y \prec z \Rightarrow x \prec z)]$, and $[x \precsim^{**} y$ if and only if $(z \prec x \Rightarrow z \prec y)]$, then $\precsim^*$ and $\precsim^{**}$ are both complete preorders. Under suitable conditions, these complete preorders can be represented by utility functions of the usual sort. Further considerations may concern the semiorder case. If $\prec$ is a semiorder and we let $[x \precsim^0 y$ if and only if $(x \precsim^* y)$ *and* $(x \precsim^{**} y)]$, then not only $\precsim^*$ and $\precsim^{**}$ are complete preorders, but also $\precsim^0$ (see Fishburn (1970)). Gensemer (1987) presented necessary and sufficient conditions for the existence of a continu-

R. Nau et al. (eds.), Economic and Environmental Risk and Uncertainty, 257–262.

ous function u representing a semiorder $\prec$ on a connected topological space (X, τ) (in the sense that $x \prec y$ if and only if $u(x) + 1 < u(y)$).

In this paper we are concerned with continuous representations of complete preorders associated with an interval order or a semiorder $\prec$ on a topological space (X, τ), which is not necessarily connected or second countable. We discuss the existence of continuous utility functions for these complete preorders, in connection with the notion of strong separability of a preference relation, which was introduced by Chateauneuf (1987).

2. Definitions and preliminary results

If $\prec$ is a *preference relation* (i.e. a binary relation which is asymmetric) on a set X, then denote with $\precsim$ and $\sim$ the *preference-indifference relation* and the *indifference relation* associated with $\prec$, that is

$$x \precsim y \Leftrightarrow not(y \prec x),$$

$$x \sim y \Leftrightarrow (x \precsim y) \text{ and } (y \precsim x).$$

Moreover, define, for every $x \in X$,

$$L_{\prec}(x) = \{y \in X : y \prec x\}, \quad U_{\prec}(x) = \{y \in X : x \prec y\}.$$

A preference relation $\prec$ is

(i) a *weak order* if $\prec$ is negatively transitive (i.e. $\precsim$ is transitive);

(ii) an *interval order* if $[(a \prec x) \text{ and } (b \prec y) \Rightarrow (a \prec y) \text{ or } (b \prec x)]$;

(iii) a *semiorder* if $\prec$ is an interval order and

$$[(a \prec x) \text{ and } (x \prec b) \Rightarrow (a \prec z) \text{ or } (z \prec b)].$$

To each preference relation $\prec$ on a set X, we may associate the following binary relations:

$$x \prec^* y \Leftrightarrow \exists \xi \in X : x \prec \xi \precsim y,$$

$$x \prec^{**} y \Leftrightarrow \exists \eta \in X : x \precsim \eta \prec y,$$

$$x \prec^0 y \Leftrightarrow (x \prec^* y) \text{ or } (x \prec^{**} y).$$

Fishburn (1970) proved that, if $\prec$ is an interval order, then $\prec^*$ and $\prec^{**}$ are weak orders (see also Bridges (1985), Chateauneuf (1987)). If in addition $\prec$ is a semiorder, then also $\prec^0$ is a weak order (see Fishburn (1970)).

From Chateauneuf (1987), a preference relation $\prec$ on a set X is said to be *strongly separable* if there exists a countable set $A\,(\subseteq X)$ such that

$$x \prec y \Rightarrow \exists a, a' \in A : x \prec a \precsim a' \prec y.$$

If $\precsim$ is a *preorder* (i.e. a binary relation which is reflexive and transitive) on a set X, then let

$$x \prec y \Leftrightarrow (x \precsim y) \text{ and } not(y \precsim x).$$

A preorder $\precsim$ is *complete* if for two elements x, $y \in X$, either $x \precsim y$ or $y \precsim x$.

If $\precsim$ is a preorder on a set X, then define, for every $x \in X$,

$$d_{\precsim}(x) = \{y \in X : y \precsim x\}, \quad i_{\precsim}(x) = \{y \in X : x \precsim y\}.$$

A preorder $\precsim$ on a topological space (X, τ) is *continuous* if $d_{\precsim}(x)$ and $i_{\precsim}(x)$ are closed sets for every $x \in X$.

A real-valued function f on a set X endowed with a preorder $\precsim$ is

(i) increasing if $[x \precsim y \Rightarrow f(x) \leq f(y)]$;

(ii) order-preserving if it is increasing and $[x \prec y \Rightarrow f(x) < f(y)]$.

A real-valued function f is a *utility function* for a complete preorder $\precsim$ on a set X if

$$x \precsim y \Leftrightarrow f(x) \leq f(y).$$

From Metha (1986), a preorder $\precsim$ on a set X is said to be *separable* if there exists a countable set $A\,(\subseteq X)$ such that

$$x \prec y \Rightarrow \exists a \in A : x \prec a \prec y.$$

Metha (1986, theorem 1) used Nachbin's separation theorem in order to find an evaluation of a continuous utility function f for a complete, continuous and separable preorder $\precsim$ on a topological space (X, τ). Metha's proof is based upon the fact that a topological space (X, τ) endowed with a complete and continuous preorder is *normally preordered* (see Nachbin (1965)). Let $\mathcal{Z}$ be a countable set such that, if $x \prec y$, then there exist z_i, z_j in $\mathcal{Z}$ with $x \prec z_i \prec z_j \prec y$. From Nachbin's separation theorem, for every z_i, z_j in $\mathcal{Z}$, with $z_i \prec z_j$, there exists a continuous increasing function $f_{i,j} : X \rightarrow [0,1]$ which separates $d_{\precsim}(z_i)$ and $i_{\precsim}(z_j)$. If we let

$$f(x) = \sum_{s \in \mathcal{S}} f_s(x) 2^{-s},$$

where $\mathcal{S}$ is a countable set whose elements are the pairs (i,j), then it is easily seen that f is a continuous utility function for $\precsim$.

From Herden (1989a), a preorder $\precsim$ on a set X is said to be *weakly separable* if there exists a countable set $A\,(\subseteq X)$ such that

$$x \prec y \Rightarrow \exists a, a' \in A : x \precsim a \prec a' \precsim y.$$

It is clear that a separable preorder is also weakly separable, while the converse is not true. Herden (1989b) observed that Metha's theorem cited above is still valid in the case when $\precsim$ is weakly separable, instead of separable. So, if $\precsim$ is a complete, continuous and weakly separable preorder on a topological space (X, τ), then there exists a continuous utility function f for $\precsim$.

3. Continuous representations

From considerations above, if $\prec$ is an interval order, then the preference-indifference relations $\precsim^*$ and $\precsim^{**}$ are both complete preorders. It is clear that, if $\prec$ is an interval order on a connected and separable topological space (X, τ), and the complete preorders $\precsim^*$ and $\precsim^{**}$ are both continuous, then, from Eilenberg's theorem, they are both represented by continuous utility functions (see Eilenberg (1941)). From Debreu's theorem, the same conclusions hold whenever (X, τ) is a second countable topological space, and $\precsim^*$ and $\precsim^{**}$ are both continuous (see Debreu (1954)).

3.1. Proposition. *Let $\prec$ be a strongly separable interval order on a topological space (X, τ). Then, if $\precsim^*$ and $\precsim^{**}$ are both continuous, there exists a pair of continuous functions u, v, which are utility functions for $\precsim^*$ and $\precsim^{**}$, respectively.*

Proof. Consider a strongly separable interval order on a topological space (X, τ), and assume that $\precsim^*$ and $\precsim^{**}$ are both continuous. From strong separability of $\prec$, it is easily seen that $\precsim^*$ and $\precsim^{**}$ are both separable. From Metha's theorem above, there exists a pair of continuous utility functions u, v for the complete preorders $\precsim^*$ and $\precsim^{**}$, respectively. So the proof is complete. $\diamond$

Chateauneuf (1987) proved that, under the assumptions of proposition 3.1, if in addition (X, τ) is topologically connected, then there exists a pair of continuous functions u, v representing $\prec$ (in the sense that $[x \prec y \Leftrightarrow u(x) < v(y)]$), where u and v are utility functions for $\precsim^*$ and $\precsim^{**}$, respectively. We don't need connectedness of the space (X, τ) whenever we are concerned only with continuous representations of $\precsim^*$ and $\precsim^{**}$.

Now let us consider the semiorder case. From considerations above, the preference-indifference relation $\precsim^0$ is a complete preorder. The following proposition relates the utility functions representing the complete preorders $\precsim^*$ and $\precsim^{**}$, respectively, to the utility function representing $\precsim^0$.

3.2. Proposition. *Let $\prec$ be a semiorder on a set X. Then, if u and v are utility functions for $\precsim^*$ and $\precsim^{**}$, respectively, $u+v$ is a utility function for $\precsim^0$.*

Proof. Let $\prec$ be a semiorder, and assume that u and v are utility functions for $\precsim^*$ and $\precsim^{**}$, respectively. First consider x, y with $x \precsim^0 y$. Then both $x \precsim^* y$ and $x \precsim^{**} y$, and it is $u(x) + v(x) \leq u(y) + v(y)$. If $y \prec^0 x$, then either $y \prec^* x$ or $y \prec^{**} x$. Observe that $y \prec^* x$ entails $u(y) < u(x)$ and $v(y) \leq v(x)$. Analogous considerations are still valid if $\precsim^*$ is replaced by $\precsim^{**}$. So $y \prec^0 x$ entails $u(y)+v(y) < u(x)+v(x)$, and the proof is complete. ◇

We omit the proof of the following corollary, which is a direct consequence of propositions 3.1 and 3.2.

3.3. Corollary. *Let $\prec$ be a strongly separable semiorder on a topological space (X, τ). Then, if $\precsim^*$ and $\precsim^{**}$ are both continuous, there exists a pair of continuous functions u, v such that u, v and $u+v$ are utility functions for $\precsim^*$, $\precsim^{**}$ and $\precsim^0$, respectively.* ◇

If $\prec$ is an interval order on a set X, then we say that $\precsim^*$ is $\precsim^{**}$ - *separable* ($\precsim^{**}$ is $\precsim^*$ - *separable*) if there exists a countable subset A (B) of X such that

$$x \prec^* y \Rightarrow \exists a, a' \in A : x \prec^{**} a \prec^* a' \prec^{**} y$$
$$(x \prec^{**} y \Rightarrow \exists b, b' \in B : x \prec^* b \prec^{**} b' \prec^* y).$$

3.4. Proposition. *Let $\prec$ be a semiorder on a topological space (X, τ), such that $\precsim^*$ and $\precsim^{**}$ are both continuous. Then, if $\precsim^*$ is $\precsim^{**}$ - separable, and $\precsim^{**}$ is $\precsim^*$ - separable, there exists a pair of continuous functions u, v, such that u, v and $u+v$ are utility functions for $\precsim^*$, $\precsim^{**}$ and $\precsim^0$, respectively.*

Proof. Let $\prec$ be a semiorder on a topological space (X, τ). If $\precsim^*$ is $\precsim^{**}$ - separable, then it is easily seen that $\precsim^*$ is weakly separable. From the slight generalization of Metha's theorem which is found at the end of section 2, there exists a continuous utility function u for $\precsim^*$. Analogously, $\precsim^{**}$ is weakly separable, and therefore there exists a continuous utility function v for $\precsim^{**}$. From proposition 3.2, $u + v$ is a utility function for $\precsim^{0}$, and the proof is complete. ◇

4. References

Bosi, G. (1993) A numerical representation of semiorders on a countable set, *Rivista di Matematica per le Scienze economiche e sociali* **16**, Fascicolo 2, 15-19.

Bridges, D.S. (1985) Representing interval orders by a single real valued function, *Journal of Economic Theory* **36**, 149-155.

Chateauneuf, A. (1987) Continuous representation of a preference relation on a connected topological space, *Journal of Mathematical Economics* **16**, 139-146.

Debreu, G. (1954) Representation of a preference ordering by a numerical function, in R. Thrall, C. Coombs and R. Davis (eds.), *Decision Processes*, New York, Wiley, pp. 159-166.

Eilenberg, S. (1941) Ordered topological spaces, *American Journal of Mathematics* **63**, 39-45.

Fishburn, P.C. (1970) *Utility theory for decision making*, Wiley, New York.

Gensemer, S.H. (1987) Continuous semiorder representations, *Journal of Mathematical Economics* **16**, 275-289.

Herden, G. (1989a) On the existence of utility functions, *Mathematical Social Sciences* **17**, 297-313.

Herden, G. (1989b) Some lifting theorems for continuous utility functions, *Mathematical Social Sciences* **18**, 119-134.

Metha, G. (1986) Existence of an order preserving function on normally preordered spaces, *Bulletin of the Australian Mathematical Society* **34**, 141-147.

Nachbin, L. (1965) *Topology and order*, D. Van Nostrand Company.

INDEX

THEORY AND DECISION LIBRARY

SERIES B: MATHEMATICAL AND STATISTICAL METHODS

Editor: H. J. Skala, *University of Paderborn, Germany*

---

1. D. Rasch and M.L. Tiku (eds.): *Robustness of Statistical Methods and Nonparametric Statistics*. 1984 ISBN 90-277-2076-2
2. J.K. Sengupta: *Stochastic Optimization and Economic Models*. 1986 ISBN 90-277-2301-X
3. J. Aczél: *A Short Course on Functional Equations*. Based upon Recent Applications to the Social Behavioral Sciences. 1987 ISBN Hb 90-277-2376-1; Pb 90-277-2377-X
4. J. Kacprzyk and S.A. Orlovski (eds.): *Optimization Models Using Fuzzy Sets and Possibility Theory*. 1987 ISBN 90-277-2492-X
5. A.K. Gupta (ed.): *Advances in Multivariate Statistical Analysis*. Pillai Memorial Volume. 1987 ISBN 90-277-2531-4
6. R. Kruse and K.D. Meyer: *Statistics with Vague Data*. 1987 ISBN 90-277-2562-4
7. J.K. Sengupta: *Applied Mathematics for Economics*. 1987 ISBN 90-277-2588-8
8. H. Bozdogan and A.K. Gupta (eds.): *Multivariate Statistical Modeling and Data Analysis*. 1987 ISBN 90-277-2592-6
9. B.R. Munier (ed.): *Risk, Decision and Rationality*. 1988 ISBN 90-277-2624-8
10. F. Seo and M. Sakawa: *Multiple Criteria Decision Analysis in Regional Planning*. Concepts, Methods and Applications. 1988 ISBN 90-277-2641-8
11. I. Vajda: *Theory of Statistical Inference and Information*. 1989 ISBN 90-277-2781-3
12. J.K. Sengupta: *Efficiency Analysis by Production Frontiers*. The Nonparametric Approach. 1989 ISBN 0-7923-0028-9
13. A. Chikán (ed.): *Progress in Decision, Utility and Risk Theory*. 1991 ISBN 0-7923-1211-2
14. S.E. Rodabaugh, E.P. Klement and U. Höhle (eds.): *Applications of Category Theory to Fuzzy Subsets*. 1992 ISBN 0-7923-1511-1
15. A. Rapoport: *Decision Theory and Decision Behaviour*. Normative and Descriptive Approaches. 1989 ISBN 0-7923-0297-4
16. A. Chikán (ed.): *Inventory Models*. 1990 ISBN 0-7923-0494-2
17. T. Bromek and E. Pleszczyńska (eds.): *Statistical Inference*. Theory and Practice. 1991 ISBN 0-7923-0718-6
18. J. Kacprzyk and M. Fedrizzi (eds.): *Multiperson Decision Making Models Using Fuzzy Sets and Possibility Theory*. 1990 ISBN 0-7923-0884-0
19. G.L. Gómez M.: *Dynamic Probabilistic Models and Social Structure*. Essays on Socioeconomic Continuity. 1992 ISBN 0-7923-1713-0

20. H. Bandemer and W. Näther: *Fuzzy Data Analysis.* 1992
ISBN 0-7923-1772-6
21. A.G. Sukharev: *Minimax Models in the Theory of Numerical Methods.* 1992
ISBN 0-7923-1821-8
22. J. Geweke (ed.): *Decision Making under Risk and Uncertainty.* New Models and Empirical Findings. 1992 ISBN 0-7923-1904-4
23. T. Kariya: *Quantitative Methods for Portfolio Analysis.* MTV Model Approach. 1993 ISBN 0-7923-2254-1
24. M.J. Panik: *Fundamentals of Convex Analysis.* Duality, Separation, Representation, and Resolution. 1993 ISBN 0-7923-2279-7
25. J.K. Sengupta: *Econometrics of Information and Efficiency.* 1993
ISBN 0-7923-2353-X
26. B.R. Munier (ed.): *Markets, Risk and Money.* Essays in Honor of Maurice Allais. 1995 ISBN 0-7923-2578-8
27. D. Denneberg: *Non-Additive Measure and Integral.* 1994
ISBN 0-7923-2840-X
28. V.L. Girko, *Statistical Analysis of Observations of Increasing Dimension.* 1995 ISBN 0-7923-2886-8
29. B.R. Munier and M.J. Machina (eds.): *Models and Experiments in Risk and Rationality.* 1994 ISBN 0-7923-3031-5
30. M. Grabisch, H.T. Nguyen and E.A. Walker: *Fundamentals of Uncertainty Calculi with Applications to Fuzzy Inference.* 1995 ISBN 0-7923-3175-3
31. D. Helbing: *Quantitative Sociodynamics.* Stochastic Methods and Models of Social Interaction Processes. 1995 ISBN 0-7923-3192-3
32. U. Höhle and E.P. Klement (eds.): *Non-Classical Logics and Their Applications to Fuzzy Subsets.* A Handbook of the Mathematical Foundations of Fuzzy Set Theory. 1995 ISBN 0-7923-3194-X
33. M. Wygralak: *Vaguely Defined Objects.* Representations, Fuzzy Sets and Nonclassical Cardinality Theory. 1996 ISBN 0-7923-3850-2
34. D. Bosq and H.T. Nguyen: *A Course in Stochastic Processes.* Stochastic Models and Statistical Inference. 1996 ISBN 0-7923-4087-6
35. R. Nau, E. Grønn, M. Machina and O. Bergland (eds.): *Economic and Environmental Risk and Uncertainty.* New Models and Methods. 1997
ISBN 0-7923-4556-8

KLUWER ACADEMIC PUBLISHERS – DORDRECHT / BOSTON / LONDON